How to Be a Scientist

Many undergraduate students choose a science degree but are not aware of how science and research work in the real world. We explain the processes of collecting, sharing and, most importantly, critical analysis of scientific research, with a focus on the life sciences. This book explains how scientific research is conceived, carried out and analysed. It outlines how research findings are constantly evolving and why that is exciting and important. Students using this textbook will learn how to design experiments, explain their data and analyse and interpret the work of others. They will learn to think about broader aspects of science, such as bias and ethics. They will gain practical skills, including understanding the use of statistical tests and how to prepare an effective presentation. Active individual and class exercises provide opportunities for students to think about difficult concepts in science and research and to include their own perspectives.

Key features of this book:

- Encourages discussion and critical thinking using individual and class exercises
- Provides real-world examples and context for difficult concepts

- Allows students to assess their understanding with practical exercises and examples
- Documents a variety of career options and opportunities from studying science
- Permits students to advocate for science with suggestions for creating and sharing research

How to Be a Scientist

Critical Thinking in the Life Sciences

Roslyn A. Kemp and
Deborah M. Brown

CRC Press
Taylor & Francis Group
Boca Raton London New York

CRC Press is an imprint of the
Taylor & Francis Group, an **informa** business

A GARLAND SCIENCE BOOK

First edition published 2024
by CRC Press
6000 Broken Sound Parkway NW, Suite 300, Boca Raton, FL 33487-2742

and by CRC Press
4 Park Square, Milton Park, Abingdon, Oxon, OX14 4RN

CRC Press is an imprint of Taylor & Francis Group, LLC

© 2024 Roslyn A. Kemp and Deborah M. Brown

Library of Congress Cataloging-in-Publication Data
Names: Kemp, Roslyn A., author. | Brown, Deborah M., author.
Title: How to be a scientist : critical thinking in the life sciences /
Roslyn A. Kemp and Deborah M. Brown.
Description: First edition. | Boca Raton, FL : CRC Press, 2024. |
Includes bibliographical references and index.
Identifiers: LCCN 2023018215 (print) | LCCN 2023018216 (ebook) |
ISBN 9780815346098 (paperback) | ISBN 9781032584690 (hardback) |
ISBN 9781003326366 (ebook)
Subjects: LCSH: Life sciences–Methodology–Textbooks. |
Life sciences–Research–Textbooks. | Biology–Methodology–Textbooks. |
Biology–Research–Textbooks.
Classification: LCC QH324 .K377 2024 (print) | LCC QH324 (ebook) |
DDC 570.72–dc23/eng/20230726
LC record available at https://lccn.loc.gov/2023018215
LC ebook record available at https://lccn.loc.gov/2023018216

ISBN: 9781032584690 (hbk)
ISBN: 9780815346098 (pbk)
ISBN: 9781003326366 (ebk)

DOI: 10.1201/9781003326366

Typeset in Garamond
by Newgen Publishing UK

For all my students at the University of Otago, especially Dr Hamish Angus, who kept my focus on education. (RK)

This book is dedicated to all my undergraduate trainees at University of Nebraska and Upstate New York who continue to inspire me to be a better educator and scientist. (DB)

Contents

Figures

Boxes

Examples

About the Authors

Deborah M. Brown, PhD, is a principal investigator at the Trudeau Institute in Upstate New York, USA, as well as an adjunct associate professor at Clarkson University. She received her PhD in Microbiology and Immunology at the University of Rochester School of Medicine and Dentistry in 2002 and completed a postdoctoral fellowship at Trudeau Institute in 2008. As an associate professor at University of Nebraska-Lincoln, she developed a passion for undergraduate and graduate education, developing curricula and instructing students in immunology, vaccine biology and professionalism before returning to Trudeau Institute in 2019 to oversee the Joint Educational Programs with Clarkson University.

With a background in cellular immunology, she has published more than 50 scientific articles on diverse topics such as anti-tumour immune responses, the role of CD4 T cells in viral infections, anti-influenza vaccine strategies and active learning techniques in the classroom. Since 2009, she has been passionate about training the next generation of scientists in the research enterprise, research methodology and scientific communication. As a member of the American Association of Immunologists (AAI) Teaching Interest Group, she has helped craft the undergraduate immunology curriculum and is a participant in the AAI active learning workshops each year. She continues to teach students in immunology and immunological techniques and supervises a unique, immersive 15-week semester-long hands-on research project for Clarkson University students investigating universal vaccines to combat influenza infection. Her work on this book represents over 20 years' experience in research and academia at a medical school, land-grant university and not-for-profit research institute and is designed to teach students to be critical thinkers as they navigate their career path in the STEM fields.

Photo credit: Photography by Dave Bull.

Professor Roslyn A. Kemp is an immunologist at the University of Otago, New Zealand. Her research focusses on immune responses in people with colorectal cancer or inflammatory bowel diseases. She graduated with a PhD in 2002 from the University of Otago and the Malaghan Institute of Medical Research and completed postdoctoral training at the Trudeau Institute, New York, USA, Oxford University, UK and the National Institute for Medical Research, London, UK, before returning to New Zealand in 2009.

Her teaching philosophy is to teach students to be scientists rather than to teach science. She has developed curricula that focus on teaching students to learn critical thinking skills and that prepare them for a variety of careers. She has also developed courses that increase the understanding of diverse world views. She is a member of the Ako Aotearoa Academy for Tertiary Teaching Excellence and has won local and national awards for teaching. She is a member of the Australia and New Zealand Society for Immunology Education Special Interest Group. She is a Council Member of the International Union of Immunological Societies and is a member of their Education and Publication Committees.

Chapter 1

The Scientific Method – An Experimental Basis for Research

Introduction and Scope

This chapter outlines the basis of experimental science as well as introduces fundamental ideas in scientific research. Some of the most common myths surrounding scientific exploration and scientists conjure up visions of white-haired people in lab coats surrounded by scraps of paper, chemicals, test tubes and experiments in progress. While not all scientific disciplines may use the scientific method (astronomy, theoretical physics, palaeontology), research in the life

DOI: 10.1201/9781003326366-1

sciences continues to generate knowledge by making observations, forming hypotheses and testing these hypotheses.

Since the subtitle of this book is *Critical Thinking in the Life Sciences*, we have limited the discussion of scientific research to those areas of biology involving basic biological research, biomedical research and clinical research. This chapter focuses on research and approaches, discussing the Scientific Method, which is often used to make observations, frame hypotheses, ask questions of basic biological phenomena and perform experiments.

Learning Objectives

- To describe the Scientific Method and compare and contrast the difference between hypothesis-driven and hypothesis-generating science.
- To appreciate how basic scientific research can be undertaken and applied to clinical problems
- To chart the timeline of the development of scientific principles throughout history
- To apply your understanding of the scientific method to real-world examples and research questions

The Philosophy of Science

There are many reasons to undertake scientific research – a desire to discover a cure for a disease or a mechanism to change how we use fossil fuels. It may also be a more basic drive to discover new information that has no obvious application. In the world of the 1850s scientist, the desire for discovery was sufficient, along with a generous benefactor, to establish a career and to build on existing information. However, commercial realities of our current world mean that the need for scientific research may be driven by the actual discovery rather than the journey. Nevertheless, the drive to discover something new and potentially important is what motivates many people who are beginning a scientific research career.

If we think of those seminal discoveries in the second half of the 19th century such as Mendel's genetic basis of heredity, Lady Mary Montague and her use of inoculation, Koch's Postulates and Pasteur's subsequent proof of the Germ Theory of Disease, we may be led to the conclusion that there is nothing left to discover. The work of Koch and Pasteur demonstrated that hypotheses could be tested by careful experimental design and interpretation, and, if these results stand up to repeated testing, then a paradigm or widely believed conceptual advance to the field is established. That is not to say that these paradigms are irrefutable, or that science does not undergo repeated refinement and adjustment. As new technologies develop, questions can be answered in different ways and new hypotheses tested to ensure that each scientific field or discipline progresses to know what has previously been unknown.

In that respect, science and scientific discovery is a process by which we continually observe the natural world, form hypotheses based on these observations and set out to test a variety of variables to prove, or, more likely, disprove the hypothesis. Hypotheses are tested by more than one scientist or group and this allows for integrity of the results, leading to a conceptual framework in which various facts and observations can be placed. For example, the basic tenant of the Germ Theory of Disease states that diseases are caused by microorganisms or pathogens that invade the body. Although the theory was first postulated in the 1700s, it took until 1840 to be able to design experiments to test this hypothesis. Three well-known scientists (Pasteur, Lister and Koch) provided the scientific evidence to prove this hypothesis, eventually leading to the identification of specific microbes that caused various diseases.

The general principle of scientific research is an adventure into the unknown – how does something work? Can I make it work better? What causes this process? How do I find a treatment for a disease? The questions themselves are less important than the fact that one is questioning.

The underlying philosophy of scientific research is simply to find stuff out.

The Scientific Method

The Scientific Method refers to an established set of processes to perform and analyse new scientific research. It begins with a question: What are you trying to find out? Then it moves to a hypothesis: What do you think is the answer to the question that you have just asked? How will you determine whether the hypothesis is true or not? What experiments can you perform to determine whether the hypothesis is true? Will the experiments give you a definitive answer; if not, is there an alternative or parallel experiment that you could perform? An experiment should determine whether the results observed agree or disagree with what was predicted in the hypothesis.

Finally, do you need to modify or refine your hypothesis based on the data that you have collected from the experiments? It is in this way that knowledge moves forward. This chapter collates the various parameters and considerations that should be incorporated into the basic scientific method.

Figure 1.1 shows the stepwise process of the Scientific Method that is a widely recognised, sensible way of approaching scientific research. It builds on principles developed as early as the 17th century to ensure accuracy and rigour in experimentation. Many students are familiar with this method – observing an effect, creating questions to potentially explains the effect, building a hypothesis, performing an experiment to test that hypothesis, analysing the data and forming a conclusion from the results. Even the standard form of scientific dissemination today, the research article (see Chapter 8), is shaped around these principles in its structure.

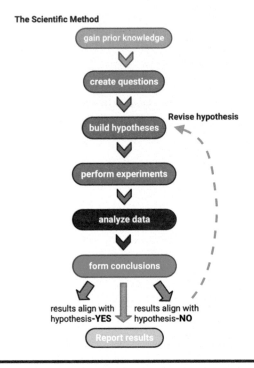

Figure 1.1   The Scientific Method. This figure outlines the scientific method for hypothesis-driven research. Often, scientists observe a particular phenomenon or effect and ask questions about that effect. A hypothesis can be generated to help explain the effect, but ultimately, experiments must be done to provide evidence for that hypothesis. After data is analysed and conclusions drawn, the results are reported or, more often, conclusions of initial experiments lead to a refinement and reworking of the hypothesis. In this way, most scientific discovery is iterative and builds on prior knowledge and results. *Figure created in BioRender.*

What Is Blood? – An Example of the Scientific Method

Early scientists created the Scientific Method to explore the world. Their experiments were less inhibited by resources or even by ethics, compared to research performed today. Most of this early scientific work was empirical and observational and was not hypothesis driven. Its goal was to first build a body of knowledge based on observations during treatments or even autopsies. For example, how did people discover blood was important for life? Humans saw the importance of blood (understanding a loss of blood is followed by a weakness of the body) but failed to apply the fundamentals of the scientific method, which in turn led them down a path of false medicine. They

1. Observed that a loss of blood was associated with weakness and eventual death
2. Incorrectly assumed blood was therefore a "life-force" inside of the human body
3. Assumed that a "life-force" must be pure, therefore, illness was due to "bad blood"
4. Invented crazy ways to remove blood from patients with no evidence to prove efficacy (leeches!)

What could have happened if they:

1. Observe that a loss of blood was associated with weakness and eventual death

2. Generate experiments to understand the nature of blood. What happens if there is:
   (a) Too little blood
   (b) Too much blood
3. Observe that blood is required for life
4. Generate hypothesis – Blood is required for life, therefore, blood from another patient should save another from death (transfusion)
5. Observe efficacy of treatment. Works sometimes, not every time, why?
6. Search for new hypothesis – Patient blood must be specific to the host in some way.

The Scientific Method is a handy mechanism to learn how to think as a scientist but is a flexible tool that can be altered in order to answer wide-ranging questions.

Types of Scientific Research

In fundamental research, approaches can be broken down into (at least) two main categories: hypothesis-driven and hypothesis-generating research. In the earlier description of the Scientific Method, the hypothesis key to the process of discovery. Hypothesis-driven research requires a hypothesis to be formed and then perhaps refined as experiments continue. However, with the advent of new technologies in the life sciences, particularly those that can generate large datasets quickly, such as whole genome analyses or single-cell sequencing, a hypothesis-generating approach is sometimes more efficient. In principle, this approach means collecting the data first and then exploring the data without a specific question in mind. Rather, one forms the question as the data reveals differences between groups or following interventions. In these situations, the experiments generally begin as hypothesis-generating and quickly become hypothesis-driven. Other studies work by collecting long-term data and making the data available to multiple researchers in different fields who can then access these data to test their own hypotheses. This approach is often the basis of large gene datasets that clinical researchers may access to determine links between gene expression and disease to validate their own clinical experimental results.

Original Research

In theory, all scientific research would be highly original and many ideas, some more creative than others, could be pursued. However, the financial limitations of research today mean that not all research is funded, and it can be difficult to determine whether something is truly original, or merely an extension of previous work. Scientific progress often works by "standing on the shoulders of giants", reproducing, refining and validating similar work performed by others (just as demonstrated with the Germ Theory of Disease). It is these extensions of previous work that sometimes can solidify paradigms, or equally important, produce paradigm shifts that change our way of thinking about previous work.

Science undergoes iterative processes to come to a consensus of proof or truth, but these journeys can be fraught with missteps and back tracking. For example, the field of immunology has seen key paradigm shifts in the last 20–30 years. In the 1970s a few immunologists hypothesised that a population of suppressive cells could put the brakes on immune responses, yet molecular and biochemical techniques in the 1980s cast doubt on this theory. Coupled with the lack of specific molecules that could identify these cells, this early work was deemed as flawed by the greater immunology community. It wasn't until the late 1990s that reports began to establish that a rare population of immune cells with high expression of a particular protein could suppress immune cell function. From that point, much work has been done on the T regulatory cell population and we now accept that these cells are important mediators of immunity, including inhibition of autoimmune disease.

Therefore, scientific work can often involve a reasonable amount of similarity with previous studies. Because scientific ideas and results are constantly reviewed by other scientists, some experiments are repeated by multiple researchers to ensure their validity. All of these reasons are important in driving discovery, but it is equally pointless to simply repeat the work of other people with no new goal.

BOX 1.1   BLUE SKIES RESEARCH

There are different reasons why scientists choose their area of study. For some, the goal is to find a new medicine or treatment for disease, or to find a way to reduce carbon emissions. However, many researchers perform very fundamental research, sometimes called blue skies research. This is work when the main goal of studying something is to find something out. It doesn't need to have a direct (or even indirect) application to real life, although this often happens eventually anyway (for example Alec Jeffreys' discovery of DNA fingerprinting (https://en.wikipedia.org/wiki/Alec_Jeffreys)). It is difficult to predict what can emerge from research until that research has been done. On top of this, the natural human drive of curiosity means that it is useful for us a species to investigate and think about things. The conversation is often around who pays for research and why. Blue skies research is hard to market to taxpayers and funders when benefits may be years away, if at all. There is an excellent blog post (https://massgap.wordpress.com/2017/12/03/the-importance-of-blue-skies-research/) reviewing the importance of blue skies research and highlighting three arguments to support investment: (1) innovation and application depends on fundamental scientific discoveries; (2) fundamental research in many disciplines often comes together unpredictably to create a new application; and (3) we currently live in a world with very serious and very complicated problems, with no straightforward way to design research to encompass it all.

Evidence and Proof, and the Importance of Conclusions

The view of many people about science is what is portrayed in the media – words like proof, evidence and data are often used interchangeably and it is sometimes difficult to interpret the meaning of scientific studies. Coupled with social media platforms, it becomes even more difficult to sort through all of the memes, shared stories and misinformation on the internet. How do you, as novice researchers and citizens, determine which media reports are based in fact and which research reports or manuscripts are legitimate? In scientific research, the burden of proof depends on the quality and type of evidence collected. *Evidence* is used to refer to the multiple building blocks that lead to *proof* – proof of a hypothesis is established when it is accepted that the evidence provided to support the hypothesis is sufficient to do so.

When designing experiments to answer a hypothesis, one must also think about the type of data collected and its value as evidence towards proving (or disproving) a hypothesis. For example, collecting reports from people about their daily diet can vary – research has shown that people are very poor at remembering what food they have eaten and often guess the amounts and times. A strategy to combat this problem is to provide your cohort with a trackable sheet or use an app with notifications to prompt them to fill in their food record. In this way, the quality of evidence is much greater, because the accuracy of the results is greater. Similarly, in laboratory-based experiments, knowing that you have measured a parameter with the same equipment each day rather than changing equipment over time means that the accuracy of the data is greater and therefore the quality of data is better. Quality data is important as evidence in order to prove a hypothesis. In hypothesis-generating research, the quality of the data collected is possibly even more important – high-volume datasets must be carefully controlled and maintained to ensure that the data hasn't been compromised.

Proof is the final result of a collection of pieces of evidence. Rarely does the proof of a hypothesis lie in one simple experiment, especially in the life sciences. Proof implies that no other conclusion could be drawn from the data collected other than the one that you have drawn. This becomes especially important when hypotheses are complex or lead on from sequential experimental results. When reviewing scientific data, peers, including students, need to determine the quality of the evidence, and decide whether it constitutes proof of the hypothesis or of the conclusions of the researchers.

Conclusions refer to the outcome and interpretation of the experimental evidence. Many researchers present work in a seminar setting and label their final slide "Conclusions" and then summarise the results of their work. This is a summary – conclusions require you to conclude something. Can you determine whether your experiment worked? Can you use your experimental data as evidence to conclude a result? The conclusion should be the final aspect of the Scientific Method – what did you find out? This conclusion will also determine the usefulness of the hypothesis. If your results align with the hypothesis, then this conclusion will be reported. If your results do not align with the hypothesis, that is an equally valid conclusion and will also be reported. This is sometimes a difficult pill to swallow for new investigators – it's okay if your results disprove your hypothesis. This then allows refinement and testing of a new hypothesis.

Summary

Science is about asking questions, designing ways to find answers to those questions and then deciding whether your answers are real. The Scientific Method is one way to develop this process; there are other methods and they vary depending on the type of science. Scientific discovery is a circular path involving cross-checking and validation by other scientists. The key skill in successful science is the ability to interpret results – yours and those of others. Good experimental design is essential, but so is the ability to critically view results and conclusions of others.

BOX 1.2 CRITICAL ANALYSIS AND TRUST

Throughout this book, we refer to critical analysis. This means that people look at data and evidence and decide if it is good quality or not. Scientists are taught critical analysis as part of their training, but many other people are not. The rise of vaccine hesitancy and misinformation on social media platforms is a current example of the discord between those who consider critical analysis essential, and those who do not or cannot. It is impossible to convince someone who believes vaccines are not effective with lots of data and graphs if they have not been trained in how to interpret graphs. For scientists, we default to trusting data, because we can understand how it is collected, how it is validated and how it isn't part of a grander conspiracy theory. However, for non-scientists, the trust is found in other places, for example from people they trust in other areas, like family or friends. It is important for scientists to be patient and to remember that they have a unique skillset. It is often much better to give up on the graphs themselves and instead focus on how a person can best learn to trust you and your data.

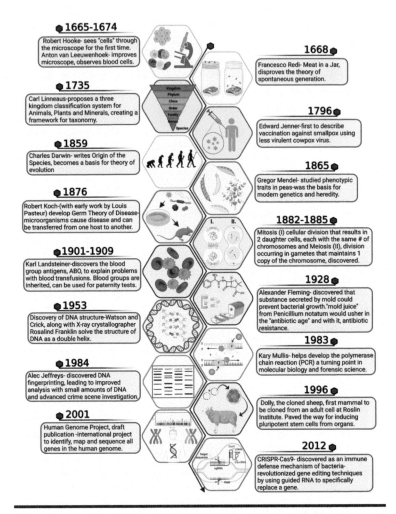

1665-1674
Robert Hooke- sees "cells" through the microscope for the first time. Anton van Leeuwenhoek- improves microscope, observes blood cells.

1668
Francesco Redi- Meat in a Jar, disproves the theory of spontaneous generation.

1735
Carl Linneaus-proposes a three kingdom classification system for Animals, Plants and Minerals, creating a framework for taxonomy.

1796
Edward Jenner-first to describe vaccination against smallpox using less virulent cowpox virus.

1859
Charles Darwin- writes Origin of the Species, becomes a basis for theory of evolution

1865
Gregor Mendel- studied phenotypic traits in peas-was the basis for modern genetics and heredity.

1876
Robert Koch-(with early work by Louis Pasteur) develop Germ Theory of Disease-microorganisms cause disease and can be transferred from one host to another.

1882-1885
Mitosis (I) cellular division that results in 2 daughter cells, each with the same # of chromosomes and Meiosis (II), division occurring in gametes that maintains 1 copy of the chromosome, discovered.

1901-1909
Karl Landsteiner-discovers the blood group antigens, ABO, to explain problems with blood transfusions. Blood groups are inherited, can be used for paternity tests.

1928
Alexander Fleming- discovered that substance secreted by mold could prevent bacterial growth."mold juice" from Penicillium notatum would usher in the "antibiotic age" and with it, antibiotic resistance.

1953
Discovery of DNA structure-Watson and Crick, along with X-ray crystallographer Rosalind Franklin solve the structure of DNA as a double helix.

1983
Kary Mullis- helps develop the polymerase chain reaction (PCR) a turning point in molecular biology and forensic science.

1984
Alec Jeffreys- discovered DNA fingerprinting, leading to improved analysis with small amounts of DNA and advanced crime scene investigation.

1996
Dolly, the cloned sheep, first mammal to be cloned from an adult cell at Roslin Institute. Paved the way for inducing pluripotent stem cells from organs.

2001
Human Genome Project, draft publication -international project to identify, map and sequence all genes in the human genome.

2012
CRISPR-Cas9- discovered as an immune defense mechanism of bacteria-revolutionized gene editing techniques by using guided RNA to specifically replace a gene.

Figure 1.2   Timeline of Scientific Discoveries in Biology.
The study of biology has roots in ancient times with Aristotle writing about and cataloguing the natural world. Similar developments were occurring in India and China that formed the basis of non-traditional medical practices. While not extensive, and limited to those from United States and Europe, the figure depicts some major biology discoveries over the last five centuries. *Figure created in BioRender.*

Example 1.1   Hypothesis-Driven Research and Hypothesis-Generating Research

Project 1: I have read that a particular type of bacteria, *Faecalibacterium prausnitzii*, can suppress inflammatory responses. I plan to use a mouse model of colitis and feed the mice extra *F. prausnitzii*. Then I will measure the amount of inflammatory mediators in the blood of the mice.

Project 2: Scientists are currently sequencing the genomes of the microbes that colonise the gut. People can send in their faeces and the scientists will quantify the types of bacteria in there.

Question 1: For each project, is this hypothesis-driven research or hypothesis-generating research or both?

Hypothesis-driven: Project 1

Hypothesis-generating: Project 2

Project 2 could also be hypothesis driven *if*, for example, the scientists proposed to see more *F. prausnitzii* in people with colitis.

Question 2: What are the possible goals of the research?

Project 1: to see if *F. prausnitz*ii affects the immune response

Project 2: to see what microbial diversity is like in a large number of people. These data could then be analysed to generate new hypotheses – is *F. prausnitzii* less prevalent in people with inflammatory diseases?

Question 3: What are the potential applications for each project?

For both projects, the research could lead to the use of probiotics to treat microbial dysbiosis, potentially in people with inflammatory gut conditions.

Question 4: What are some of the limitations in what one can conclude from these experiments?

Project 1: Is the result relevant in humans? Is it because *F. prausnitzii* directly affects inflammatory responses, or due to changes in microbial diversity?

Project 2: With so much data collected, how can you tell if there is a direct effect of one type of bacteria? How do you distinguish a real effect from the background "noise" of all the other bacteria, and the differences in individual people?

Example 1.2   Capitalising on Serendipity –
The Penicillin Story

There are many examples of serendipity in science – making an unexpected discovery when conducting experiments, or having an experiment turn out opposite to your predictions due to unforeseen circumstances. In fact, author Isaac Asimov said, "The most exciting phrase to hear in science, the one that heralds new discoveries, is not 'Eureka!', but 'That's funny…'" That's not to imply that serendipitous discoveries are mere accidents or the result of chance happenings. The ability to see that an unexpected result may be significant is also attributed to the wisdom and critical thinking skills of scientists leading those experiments.

One such story of a serendipitous discovery comes from Alexander Fleming in 1928 (see Figure 1.2). Fleming (1881–1955) was a physician/scientist and bacteriologist by training in London working on the bacteria *Staphylococcus aureus,* when he noticed that there was mould growing in one of his bacterial plates. He had different strains of Staphylococci bacteria (including *Staphylococcus aureus*) and allowed them to grow at room temperature, instead of the normal 37°C temperature. As he was going on holiday for several weeks, he left them on the bench top to see if they could grow at the lower temperature. When he returned, he saw that the Staphylococci on the plates had grown. He observed that a contaminating mould also grew on one of the plates. There was a defined area around the mould where the bacteria did not grow, while areas further from the mould had considerable bacterial growth. He hypothesised that the mould secreted something that killed the bacteria. Fleming cultured the mould in a nutrient broth and showed that the broth from the mould ("mould juice") inhibited the bacterial growth. He published the results in 1929 and called the substance penicillin. However, the publication received little attention and Fleming couldn't purify the substance without making it inactive.

Enter Drs Howard Florey and Ernst Chain at the University of Oxford. A decade went by before pathologist Dr. Howard Florey and biochemist Dr. Ernst Chain were studying the effects of an enzyme, lysozyme, when they came across the paper describing penicillin. Florey, Chain and colleagues at the University of Oxford developed a method to isolate and purify penicillin from the culture broth without destroying its anti-bacterial properties. This development meant penicillin could be produced in large quantities. But it still required close to 500 litres of mould broth to produce enough drug to study it. Florey first used the drug to demonstrate that treatment of rats with penicillin cured them of a lethal dose of the bacteria, Streptococcus. While World War II was raging in Europe, the need to provide enough of the drug and to test in people was crucial as this wonder drug could be used for battlefield infections. Conducting biomedical research during war was difficult and it was even suggested that all in Dr. Florey's group put the penicillin spores on their clothing so anyone who survived a German attack would be able to carry on the work.

Large-scale production moves to the United States: Dr. Florey began testing the drug on human patients but the limited supply of purified penicillin was running out, and several drug companies in Britain denied his request to begin large-scale production. In 1941, Dr. Florey and Norman Heatley contacted an expert in penicillin mould at the Department of Agriculture, USA. Large-scale production for penicillin began. Armed with significantly better fermentation media and a penicillin mould strain that grew quickly, four top drug companies (Merck, Squib, Pfizer and Lederle) began the industrial manufacturing of penicillin. By March 1942, the U.S.-manufactured penicillin was given to the first patients and by 1943, the War Production Board increased the production of penicillin to be used on the battlefield.

Penicillin was a major scientific discovery that heralded the antibiotic age. Although discussed as a serendipitous finding, it clearly was no accident that Fleming hypothesised that a substance in the mould may be actively inhibiting and killing certain pathogenic bacteria. The collaborative wisdom of the Oxford group found a way to purify the penicillin without endangering its potency. Finally, the war effort confirmed that penicillin was a miracle drug that showed promise as a therapeutic for a wide range of infections. Drs Fleming, Florey and Chain were awarded the Nobel Prize in Physiology or Medicine in 1945.

Potential Careers

Students who complete a science degree don't have to have a career working in a lab! The skills one acquires as a science student are applicable to many work situations. There are also thousands of science-adjacent jobs, where the employee is working in science and contributing to science and research. In each chapter we will feature a different career as an example of where a science degree can take you.

International Science Organisations

Science is collaborative and international – scientists must work together and be organised. There are several international organisations that coordinate scientific activities, prepare position statements and provide advice to the public. High-level examples include the World Health Organisation (WHO) and the International Science Council (ISC). People with science degrees are sought after in these organisations to work in areas developing policies around science communication, gender equity in science, development of resources for doctors or the public and coordinating membership of the organisations. For examples of current vacancies and testimonials from employees in such societies, see here: www.who.int/careers and here: https://council.science/about-us/jobs/.

Resources and Further Reading

Resources Used in This Chapter

■ The Scientific Method
www.sciencebuddies.org/science-fair-projects/
science-fair/steps-of-the-scientific-method
■ Blue Skies Research
https://massgap.wordpress.com/2017/12/03/the-
importance-of-blue-skies-research/

Linden B. Basic Blue Skies Research in the UK: Are
we losing out? *J Biomed Discov Collab*. 2008 Feb
29;3:3. doi: 10.1186/1747-5333-3-3. PMID: 18312612;
PMCID: PMC2292148.
   A note on Alec Jeffreys: https://en.wikipedia.org/
wiki/Alec_Jeffreys

■ Retraction of Wakefield paper claiming a link
between autism and MMR vaccination

Retraction – Ileal-lymphoid-nodular hyperplasia,
non-specific colitis, and pervasive developmental
disorder in children. *Lancet*. 2010 Feb 6;375(9713):445.
doi: 10.1016/S0140-6736(10)60175-4.

Additional Websites That Might Be Useful

■ Philosophy of science
https://blogs.scientificamerican.com/doing-good-
science/pub-style-science-philosophy-hypotheses-
and-the-scientific-method/

- Serendipitous discoveries
  https://owlcation.com/stem/Serendipity-The-Role-
  of-Chance-in-Making-Scientific-Discoveries

Copeland S. On serendipity in science: discovery at
the intersection of chance and wisdom. *Synthese*.
196, 2385–2406 (2019). https://doi.org/10.1007/s11
229-017-1544-3

- Distrust of science
  www.aamc.org/news-insights/why-do-so-many-
  americans-distrust-science
  www.newyorker.com/news/news-desk/the-mistr
  ust-of-science

Research Publications for Further Reading

1. "Good" science and "bad" science

Parsons ECM, Wright AJ. The good, the bad and
the ugly science: examples from the marine science
arena. *Front Mar Sci*. 2015;2:33. doi: 10.3389/
fmars.2015.00033.
 Brumfiel, G. Controversial research: Good science
bad science. *Nature*. 2012;484:432–434. https://doi.
org/10.1038/484432a

2. Role of serendipity in drug discovery

Ban TA. The role of serendipity in drug discovery.
Dialogues Clin Neurosci. 2006;8(3):335–44.
doi: 10.31887/DCNS.2006.8.3/tban.

3. Faecalibacterium and immune responses

Sokol H, Pigneur B, Watterlot L, Lakhdari O, Bermúdez-Humarán LG, Gratadoux JJ, Blugeon S, Bridonneau C, Furet JP, Corthier G, Grangette C, Vasquez N, Pochart P, Trugnan G, Thomas G, Blottière HM, Doré J, Marteau P, Seksik P, Langella P. *Faecalibacterium prausnitzii* is an anti-inflammatory commensal bacterium identified by gut microbiota analysis of Crohn disease patients. *Proc Natl Acad Sci U S A*. 2008 Oct 28;105(43):16731–16736. doi: 10.1073/pnas.0804812105.

Exercises

Individuals

Exercise 1

I hypothesise that eating a high-fat diet leads to weight gain. I have collected food diaries from a large cohort of people and monitored their weight over a year. Analysis of the results demonstrated an association between consumption of chips and more weight gain.

   a. How could the experimental design be improved to increase the quality of evidence?
   b. What are the conclusions that can be made based on this experiment? What conclusions should not be made?
   c. Have I generated proof for my hypothesis?

Exercise 2

In 1998, Andrew Wakefield and colleagues published an article in the *Lancet* that sparked a case of scientific fraud that has had long-ranging implications for childhood vaccination and autism. Although that paper was ultimately retracted (Retraction – Ileal-lymphoid-nodular hyperplasia, non-specific colitis, and pervasive developmental disorder in children. *Lancet.* 2010 Feb 6;375(9713):445. doi: 10.1016/S0140-6736(10)60175-4), the impact of those findings is still being felt today with overall vaccine hesitancy in the population. Looking at the paper and findings, some key phrases and ideas emerge. In the authors' Interpretation section: "We identified associated gastrointestinal disease and developmental regression in a group of previously normal children, which was generally associated in time with possible environmental triggers." Another important statement in the methods reads: "Onset of behavioural symptoms was associated, by the parents, with measles, mumps, and rubella vaccination in eight of the 12 children, with measles infection in one child, and otitis media in another. All 12 children had intestinal abnormalities, ranging from lymphoid nodular hyperplasia to aphthoid ulceration."

    a. Given these statements, what evidence is there in this paper that there is an association between developmental regression and environmental triggers? Which environmental triggers are the authors describing? How were these environmental triggers measured?

b. Is a cohort of 12 children sufficient to demonstrate association? Why or why not? What statistical test was used to assess association?

c. Table 1 shows all children in the study already had a diagnosis of autism, so any other symptoms or results would automatically associate with the developmental regression. What other controls should the authors have used to compare with their study cohort?

Class

Exercise 3

Design both a hypothesis-driven and a hypothesis-generating research project on the same topic. Compare and contrast the benefits and limitations of each technique.

Which technique would you choose for your topic, why?

Exercise 4

In clinical trials, subjects were vaccinated, then boosted three weeks later, with either Pfizer/BioNtech SARS CoV-2 vaccine or a saline placebo. Blood was collected prior to vaccination, three weeks after the first vaccine and three weeks after the second vaccine to assess antibody levels against SARS CoV-2. Neutralising antibody titres were also analysed and showed that the two-dose vaccine regimen induced antibodies that could prevent infection to a higher level than antibodies after one vaccine dose.

   a. What are the conclusions that can be made with regard to antibody responses after vaccination?

   b. Does this study prove that antibody responses are responsible for protection? Why or why not? Could you design a clinical trial experiment to prove that antibody responses are responsible for protection?

      In the follow-up study, approximately 40,000 participants were recruited with half of subjects receiving two doses of vaccine and the other half of subjects receiving placebo. Studies show that the group receiving vaccine were protected from severe COVID-19 disease compared to placebo groups.

   c. Do these studies prove that vaccination with Pfizer SARS CoV-2 vaccine prevents infection with the virus?

   d. How can the study design be improved to demonstrate that the vaccine can inhibit infection, what tests would be done to show this?

Useful resources for this exercise:
   www.vet.cornell.edu/news/20210329/how-vacci nes-work
   www.youtube.com/watch?v=jeN8v5I5VNA

Walsh EE, Frenck RW Jr, Falsey AR, Kitchin N, Absalon J, Gurtman A, Lockhart S, Neuzil K, Mulligan MJ, Bailey R, Swanson KA, Li P, Koury K, Kalina W, Cooper D, Fontes-Garfias C, Shi PY, Türeci

Ö, Tompkins KR, Lyke KE, Raabe V, Dormitzer PR, Jansen KU, Şahin U, Gruber WC. Safety and immunogenicity of two RNA-based Covid-19 vaccine candidates. *N Engl J Med.* 2020 Dec 17;383(25):2439–2450. doi: 10.1056/NEJMoa2027906.

Thomas SJ, Moreira ED Jr, Kitchin N, Absalon J, Gurtman A, Lockhart S, Perez JL, Pérez Marc G, Polack FP, Zerbini C, Bailey R, Swanson KA, Xu X, Roychoudhury S, Koury K, Bouguermouh S, Kalina WV, Cooper D, Frenck RW Jr, Hammitt LL, Türeci Ö, Nell H, Schaefer A, Ünal S, Yang Q, Liberator P, Tresnan DB, Mather S, Dormitzer PR, Şahin U, Gruber WC, Jansen KU; C4591001 Clinical Trial Group. Safety and efficacy of the BNT162b2 mRNA Covid-19 vaccine through 6 months. *N Engl J Med.* 2021 Nov 4;385(19):1761–1773. doi: 10.1056/NEJMoa2110345.

Chapter 2

The Scientific Process – Planning, Thinking, Doing

Introduction and Scope

What are the tools that scientists use to formulate their ideas, figure out whether they're good or not and then apply to funding agencies with a proposed research project? Good ideas don't just drop out of the sky. A stepwise process from reading the background literature, to identifying gaps in our knowledge, to proposing a series of well-controlled experiments that aim to solve a significant problem, to reporting these results to the scientific community and general public, has to be followed.

DOI: 10.1201/9781003326366-2

Scientific research is a mix of creativity and rigour. New ideas need to be explored in a manner that ensures reproducible and accurate results. This chapter explores the pathways from a new idea, including how that idea came about, through to publication of results and uptake by the medical or research community. Later chapters cover the specifics of some of the steps along the pathway.

This chapter introduces you to basic concepts around designing and carrying out a scientific project.

Learning Objectives

- To identify the characteristics of a rigorous experimental design
- To compare and contrast the different options for reporting results and identify stakeholders
- To appreciate the strengths of your research team and propose ways to expand expertise

Coming Up with an Exciting Idea

The success of your scientific project depends on acquiring funding to support it, and on the peer review process that allows for publishing. Both of these processes require the researcher to sell their idea to a general, or perhaps very specialised, audience. Because funds are limited, and research publications are sometimes considered a measure of quality of research (controversial), reviewers look for many things, including the likelihood of success and your proven ability within your field of research. However, the most important thing they look for is novelty. How does your proposal build on existing knowledge? But how do you come up with a new idea and what defines new? Is it testing a known protein in a new disease model? Is it comparing responses in two different disease models? Is it the crazy unproven idea with no preliminary data to support the hypothesis? It could be any one of these. So how do you find a new idea?

It is important to be aware of your own abilities and expertise. While it is good to expand your own knowledge, your real strength lies in the area in which you are trained. Having said that, collaboration and interdisciplinary studies are becoming more important. Within your research area, what other expertise might be useful? Who could you work with – a surgeon, a statistician, a protein chemist or an engineer? Building a research team to test new ideas and to expand the techniques and analyses available are all good approaches.

It is also essential to be able to identify the gaps in knowledge. What is missing in your field? What are the unanswered questions? How could you address them? What do you wish someone had already discovered? It is important to be aware of old research, too – many discoveries that actually had been made before have, in fact, been rediscovered using new and exciting tools. Reading published papers is essential to see what is and isn't already known, and what others have identified as possible future areas of research (Box 2.1). New technologies are another way to generate new ideas. It is possible that existing research could be enhanced or improved by collecting higher quality data using new techniques or equipment.

Designing a Research Project

Depending on the goal of the project (for example, as a grant funding application), there are several key aspects of which to be aware. The first is the hypothesis (discussed in Chapter 1). The second is the goal of the research – all research projects need an overall aim, for example, to cure cancer. The overall aim explains where you see your research project is ultimately leading to, but it is not necessarily a specific goal that you will achieve in the short term. The next level are the objectives. These are the "mini-goals" of your project – what specifically are you trying to achieve? For example, (i) to identify the protein that is mutated in cancer; (ii) to modify the protein so that the mutation is fixed; (iii) to show that the modified protein cures cancer. Within these objectives is the chance to explain the methodology and technology you will use to address specific questions.

The timeframe in which you can achieve your research goals is also important. You need to be realistic about what can and cannot be done with the resources you have or hope to have. You may need to depend on recruitment of patients, or breeding of animals, or long-term delivery of reagents. These timeframes are also dictated somewhat by different types of grant proposals, or different amounts of time that the researcher can spend on a given project. For example, funding sources can be over a two-year (for example, USA NIH R21 proposals) or five-year (USA NIH R01 proposals) time frame and the trap that many fall into is proposing work that is too ambitious for a short-term grant.

The significance or potential application of your research is often difficult to write without overstating your outcomes. However, in a space with limited funding, you need to be able to provide a reason why the funders should choose your project over others. This becomes important when considering funding applications to charities compared to government or institutional funding – the latter may be happy to fund science per se, whereas charities need to be accountable to their donors who have given money expecting to see a particular result. Tied into this is an expectation from taxpayers, who fund the government, that high quality and relevant research is being done with their money. While it is true that you may not cure a disease, it is good to be able to show a pathway from your expected research outcomes to how others could use these to develop other areas of research that would ultimately lead to some change in practice, management or treatment, as well as to health outcomes.

Requirements for the level of detail of the methodology vary, but it is important to have an experimental plan in place – this may involve using established techniques to answer your research question, or as part of the proposal, to develop new approaches or modify existing techniques. The type of methodology to be used depends on the funding and resources available, as well as the expertise. Any proposal without a clear experimental plan (Chapter 3) is unlikely to be successful – note that this includes a plan for the acquisition and analysis of the expected data (Chapter 5) – do you have the software and expertise to analyse the data properly? Have you thought about statistical analyses and how that may affect your research design (Chapter 4), for example, recruitment of patients for a clinical trial? How reliable are your measures? Building on the data published by others is particularly useful at this stage.

Collaboration is a key component of research. Most publications involve research from different institutions that share ideas, resources, expertise and technology. The ability to generate large datasets with new technologies means that multiple researchers can contribute to the interpretation of these data. A good research proposal acknowledges the need for support from others, and also outlines a plan for how the overall skillset will be used to achieve the goals. For example, do you need a clinician to be involved in patient recruitment and data collection? Do you need a statistician to contribute to analysis of data? Do you need expertise in both genomics and biochemistry to address your questions?

BOX 2.1   HOW TO READ A PAPER

At first, reading a paper can be daunting – the formatting is usually unfriendly, the figures small and squashed and the jargon sometimes overwhelming. However, the key is to focus on the overall picture and message. Details and critique can come in a second read.

Begin with the Title – There should be enough detail in a title to pique interest and to confirm relevance. Ideally, a title tells you the result or the novel finding. This allows you to have a general idea of what is going to be explained.

Introduction – The introduction should give an overview of the current status of the field of research and identify the gap in the knowledge that the paper addresses. An introduction may also provide a short summary of what the experiments were and what the results were. The introduction is a great source of references to use to understand the field.

Figures and Results – Paper figures and legends are the results of the experiment. Each figure legend title and content should explain in sufficient detail what was done and also what was found. The results section of a paper can be read in conjunction with each figure – often this section is broken down into subsections for each figure and so you can look at the picture, then find the associated text. For each figure, you should be able to (i) identify what question they are asking, (ii) understand the experimental design, (iii) determine what the answer to the question is and (iv) identify the next step or experiment that should be done to address the next part of the overall research gap.

Discussion Paragraph 1 – A discussion section usually has a summary paragraph first that recaps the research question, the research findings and a comment on the significance of the research.

At this stage, you should have a solid understanding of what the paper discovered and why it is important. A second read can then include:

The Rest of the Discussion – This puts the research into a broader context and identifies limitations and strengths of the approach. It may also identify which questions need to be answered next.

Abstract – These are useful for initially finding a paper but are often so short and dense that they are overwhelming and confusing when first reading a paper. However, an abstract provides a summary of the rationale, methods, results and significance, and so is useful to read after earlier steps to help you determine whether you agree with the authors' findings.

Methods – These can be referred to earlier if needed for clarity. Methods sections are designed to provide enough information that the experiments could be repeated by someone else, but they are often very concise in a paper format. Understanding complex experiments can be difficult, and so focusing on the result is more important. However, it is also important to understand experimental design and to critique it if necessary.

Acquiring Funding

Most science projects are funded by a competitive grant process – the details of these are covered in Chapter 7. However, it is impossible to separate the scientific proposal design from the requirements and restrictions of individual funding bodies. When designing aims, hypotheses, methodology and, particularly, the scope of the project, then funding must be considered – how long are the grants going to supply funds? Do they fund translational or fundamental research? Do they fund salaries for people to actually perform the experiments? It is not unusual for a research proposal to be sent to multiple agencies or to be broken into component parts for funding.

Reporting

The final aspect of the scientific process is reporting the results and conclusions. This is covered in more detail in Chapter 5, but the audiences most likely to be targeted are:

1. The research-field-specific scientific audiences, by publication in peer-reviewed scientific journals. Publication at this level ensures that your work is judged and read by researchers who know a lot about your research area and are likely to be most interested in it. The work may also be presented at a scientific conference,

which reaches the same audience but is not held to the same level of quality – conferences are therefore a good place to sound out new ideas or test conclusions in front of your peers. The use of preprint servers (where research is publicly available before it has been peer reviewed and officially published – for example, bioRixv) provides a faster way to reach the scientific audience but have less rigour than the conventional pathway. The rapid dissemination of research results during the SARS CoV-2 pandemic has demonstrated the advantages and importance of preprint servers.

2. Those likely to see application in your work, for example, doctors who want to implement your findings into clinical practice or pharmaceutical companies who may want to develop drugs based on your work. These stakeholders may read your research in scientific settings but reports or seminars to these audiences are also common, especially if some of the research funding has come from these groups. Another example is government departments, ministries or agencies that may include the data in relevant policies or strategic documents.

3. Those invested in your research need to know the outcomes of their investment and be able to report this to their stakeholders. This may be in its simplest form a short report to the funding agency stating that the proposed objectives were achieved. Given that research fields and technologies change quickly, and

sometimes research projects do not give the expected outcomes, some justification or explanation of the project progress will also be required. This is common in reporting to national funding agencies such as National Institutes of Health (NIH) in the United States, Medical Research Council (MRC) in the United Kingdom and Health Research Council (HRC) in New Zealand.

4. While the public are often interested in scientific findings, particularly when the majority of funding may ultimately come from taxpayer funds, explaining significance and achievement to a general audience can be difficult. Strong and trustworthy relationships with the media are vital in order for your research to be reported accurately, and many institutions have a media office in order to help you frame your results appropriately. Science communication aims to teach science faculty and other research scientists better ways to communicate with the general public. The hope of many of these organisations is to increase trust in scientists and their findings as well as improving overall science literacy in the public.

The use of social media platforms, such as Twitter, widens the reach both to scientific and lay audiences – social media helps to immediately publicise new findings and generate a buzz about new research. Funders and research institutes are recognising the informal impact of a social media presence as a means to improve the value of research, as well as more conventional metrics.

BOX 2.2 HOW TO COME UP WITH NEW IDEAS

Non-scientists often misunderstand how science works. It is easy to see how a philosophy degree (a PhD) exists in a field with clearly lots of reading and thinking. However, science PhDs are also philosophy degrees. They involve huge amounts of reading and thinking to generate new ideas and then testing those ideas.

Competitive cooking shows, like MasterChef, often feature contestants talking about the dessert course as "science" because there are very detailed steps to follow, in contrast to other courses, where creativity and winging it are common. This could not be further from the truth – the dessert is like following an experimental protocol, but science involves creativity, spontaneous decisions, reactivity to new situations and a basic background knowledge – just like cooking. Also, closed mussels are fine to eat (www.abc.net.au/science/artic les/2008/10/29/2404364.htm).

Ideas on how to be creative in science:

1. Write the idea down, to think more deeply about it, now and later.
2. Talk to other people about it – different approaches from different people are the basis of successful interdisciplinary research.

3. Cultivate your own body of knowledge, read around similar topics, or old articles, to try to bring a unique perspective to your body of knowledge.
4. Study good ideas of other scientists.
5. Rigorously approach and examine existing assumptions on which your idea is based.

Summary

The development or discovery of new research findings is a pathway with multiple steps. Each step is subject to review by other scientists (usually for free!) so that quality is assured, and that data can be validated and lead to even more new ideas. Science is very much a community of volunteers, working together to advance knowledge.

Example 2.1   Sense about Science

Sense about Science (senseaboutscience.org) is an independent non-governmental organisation that promotes evidence and science to the public. Their website provides information on current science topics – these resources are free to download and share. They also present evidence addressing scientific claims in the traditional media as well as on social media. Their hashtag #AskForEvidence helps the non-scientific public work out what is true or false, and more importantly, *why* the claim is true or false. They also provide extensive resources on topics such as understanding risk.

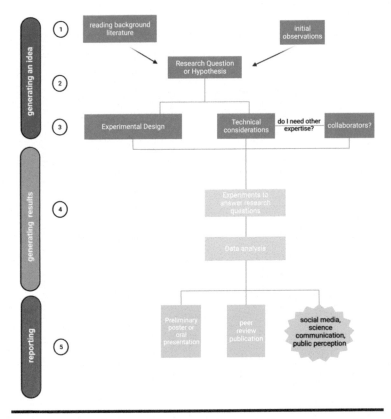

Figure 2.1 Pathway of Design to Output. Generating an idea is only the first part of the scientific process. Forethought and learning the background of a research project is needed before a research question or hypothesis can be formulated. An initial observation often focuses the research question. From the research question/hypothesis, you begin to build the experimental design, that is, what experiments and techniques are you going to use to generate data that support your hypothesis? Do you need particular expertise from a collaborator? Is there a new technique that can be used to answer your questions? Once an experimental plan is developed, the experiments are completed and interpretable results are generated. Chapters 3 and 4 discuss more considerations for the types of controls

and statistical analyses needed to adequately interpret data generated. Chapters 5 and 8 will discuss the basics of reporting data and publishing manuscripts. *Figure created in BioRender.*

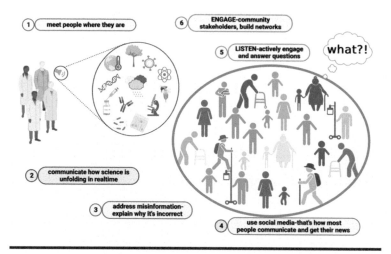

Figure 2.2   Communicating Science to Non-Scientists.
As scientists, we need to bridge the gap in knowledge and communicate in real time in an inclusive and nonpartisan manner. To communicate effectively with a varied public, you must be clear in your message and acknowledge that science is dynamic, and our understanding of the data can shift as more information becomes available. It's also important to address misinformation and explain why it's incorrect – a lot of confusion around science messaging stems from misconceptions and misunderstanding the interpretation or the significance of the data. It's crucial that scientists *engage* and *listen* to the public; this includes community stakeholders such as local government and educational networks. Use social media to get your message across. That's how most people get their information on a topic. Social media platforms can also be used to address the comments and questions people have surrounding scientific discoveries. *Figure created in BioRender.*

Example 2.2   Science, Quickly

Science, Quickly podcast is a product from *Scientific American*. They publish a regular podcast, summarising recent scientific findings in under ten minutes, with a fun perspective. Since 2020, they have also included COVID Quickly – a regular summary of COVID-19-related science. These are ideal communication tools to reach the public in an accessible way – we all listen to podcasts while driving, running or cooking, for example.

www.scientificamerican.com/podcast/60-second-science/

Potential Careers

Many science careers don't require specialisation; in fact, some are better for science generalists, people who like all science. Many people like science research and want to be involved but have decided that a research career is not for them.

Research Advisors

Research advisors are people who coordinate and review grant applications for an institution. These people are essential, not only for handling the administrative requirements of grant funding submissions but also to provide feedback on content, scope and whether the grant makes sense. Because research advisors often must provide feedback on a variety of grant types and in different disciplines, it is an ideal job for someone who is interested in all areas of science, rather than in one specialised topic. Research advisors are excellent communicators, and require good writing skills. They also need to be highly organised and not afraid to keep researchers to deadlines.

Resources and Further Reading

Resources Used in This Chapter

■ Science communication
www.scientificamerican.com/podcast/60-second-science/
www.senseaboutscience.org

Additional Websites That Might Be Useful

■ Creativity, experimental design and how to formulate research questions
www.ibiology.org/professional-development/creativity-in-science/
■ The preprint revolution

The COVID-19 pandemic has highlighted the importance of rapid dissemination of research results to the scientific community. Here are three links that discuss the rise of the preprint server and its positive impact on reporting results and reviewing results.
www.ibiology.org/science-and-society/preprint/

www.the-scientist.com/news-opinion/opinion-the-rise-of-preprints-is-no-cause-for-alarm-68667?utm_campaign=TS_DAILY_NEWSLETTER_2021&utm_medium=email&_hsmi=121557939&_hsenc=p2AN qtz-_t59MwlcmIRceMUsZJns4O9UPSRDWxwNBoF9_mRo0eskuBGsZKNrU4X9B_j4glBUA8kSHbVmcbsS8 P_3-Z7L2SDbYr6igw5b18Tq2uvE0jqBbnfy0&utm_cont ent=121557939&utm_source=hs_email
www.nature.com/articles/d41586-020-03564-y

- ■ Science communication
  www.aldacenter.org/
- ■ Understanding evidence
  https://askforevidence.org
- ■ Research design
  www.surveymonkey.com/market-research/resour ces/steps-experimental-research-design/

Research Publications for Further Reading

1. Research design

Noordzij M, Dekker FW, Zoccali C, Jager KJ. Study designs in clinical research. *Nephron Clin Pract.* 2009;113(3):c218–c221. doi: 10.1159/000235610.

Mendoza AE, Yeh DD. Study designs in clinical research. *Surg Infect* (Larchmt). 2021 Aug;22(6):640–645. doi: 10.1089/sur.2020.469.

2. Science communication and understanding evidence

Gustafson A, Rice RE. A review of the effects of uncertainty in public science communication. *Public Underst Sci.* 2020 Aug;29(6):614–633. doi: 10.1177/0963662520942122.

3. The American Gut Project

McDonald D, Hyde E, Debelius JW, Morton JT, Gonzalez A, Ackermann G, Aksenov AA, Behsaz B, Brennan C, Chen Y, DeRight Goldasich L, Dorrestein PC, Dunn RR, Fahimipour AK, Gaffney J, Gilbert JA, Gogul G, Green JL, Hugenholtz P, Humphrey G, Huttenhower C, Jackson MA, Janssen S, Jeste DV, Jiang L, Kelley ST, Knights D, Kosciolek T, Ladau J, Leach J, Marotz C, Meleshko D, Melnik AV, Metcalf JL, Mohimani H, Montassier E, Navas-Molina J, Nguyen TT, Peddada S, Pevzner P, Pollard KS, Rahnavard G, Robbins-Pianka A, Sangwan N, Shorenstein J, Smarr L, Song SJ, Spector T, Swafford AD, Thackray VG, Thompson LR, Tripathi A, Vázquez-Baeza Y, Vrbanac A, Wischmeyer P, Wolfe E, Zhu Q; American Gut Consortium; Knight R. American Gut: an Open Platform for Citizen Science Microbiome Research. *mSystems.* 2018 May 15;3(3):e00031–e00018. doi: 10.1128/mSystems.00031-18.

Exercises

Individual

Exercise 1

Choose a recent science story from the popular media. Critique the original scientific study and then discuss the accuracy of the media reporting of the study.

Prepare a one- to five-minute podcast describing the finding and significance.

Exercise 2

You have identified a new technology that might help answer your research question, but your lab doesn't have expertise in that technology. Outline ways in which you can expand your expertise (with collaborators, research resources, etc.) to be able to apply new technologies to your research.

Class

Exercise 3

Using findings published from the American Gut Project (www.mymicrobiome.info/en/news-read ing/the-american-gut-project), design a one-page research proposal to investigate one of the findings further. Include an Aim, Hypothesis, three Research Objectives, an outline of Methodology and a Statement of Significance.

Exercise 4

Identify a product in the public realm – through advertising or social media – that states a scientific claim for efficacy or importance. This may be a food supplement, a new treatment or a claim about vaccines – there will be many to choose from!

Using the guidelines from Sense about Science, including their information at https://askforevide nce.org/, contact the people advertising the product and ask for the evidence for their claim. Then make a video or animation explaining your findings –for example, see https://askforevidence.org/articles/ani mations

Chapter 3

Scientific Controls

Introduction and Scope

The purpose of this chapter is to introduce you to different types of controls (experimental, statistical, technical); to provide guidance in choosing the appropriate controls and to discuss reporting of controls when publishing. This is an extensive topic, so we have focused mostly on *why* controls are important. Other chapters address controls in the context of other concepts.

Learning Objectives

- To define and identify positive and negative controls within experiments
- To understand the difference between technical replicates and biological replicates

 DOI: 10.1201/9781003326366-3

■ To design an experiment using technical and biological replicates as well as positive and negative controls

Controlling Experimental Design

Experimental design involves formulating a question and trying to find the answer. The value in a good experiment is being able to determine what is, and is not, a valid finding. This concept is the key to scientific controls. Is your result negative because your answer is no or because you didn't detect the thing you were looking for? What constitutes a positive result? Do you have a sample that you know will show an increase in the assay you are performing?

An experiment ideally contains a positive control and a negative control, alongside the test groups. This cannot always be easily done. A positive control is an experimental condition that allows you to *definitely* detect the change you are looking for. A negative control is an experimental condition that allows you to determine a baseline or background reading. Some experiments may contain only a negative control, for example, *in vivo* manipulations of animals. In these situations, a true positive control may not be possible; however, the same concept can be integrated by using an experimental condition that has worked previously and been accepted in the published literature. An ideal negative control for animal or cell experiments is one which is not capable of doing the thing that you are measuring, for example, using a knockout mouse or a cell engineered to be unable to produce a molecule. These are excellent controls but often difficult to include in every experiment (and must also be interpreted with the caveat that gene knockout models may have off target effects). In some experiments, multiple negative and positive controls need to be included to correctly determine an effect.

Tumour Immune Responses

i.   Measuring a Tumour Immune Response

Question: Do tumour cells cause T cells to proliferate?

In order to test whether T cells respond to a tumour, the T cells and tumour cells are cultured together in the lab. Since T cells proliferate when they are activated by a tumour, the experimental readout is T cell proliferation.

The *negative control* is the addition of something that will definitely *not* make the T cells proliferate, for example, media only.

The *positive control* is the addition of something that will definitely make the T cells proliferate so you can measure proliferation (for example, stimulation with polyclonal antibodies that activate all T cells).

In addition, it may be wise to include an *additional negative control* that consists of tumours cultured alone (media only) to ensure that the tumour cells aren't undergoing proliferation on their own, which could confound the experimental readout.

ii. Measuring the Efficacy of a Tumour Vaccine In Vivo *in Mice*

Question: Does an administered vaccine reduce tumour growth *in vivo*?

In order to test whether the vaccine works, it is administered to a group of mice. The mice are then injected with tumour cells and the tumour growth is measured. The experimental readout is tumour size.

The *negative control* is no vaccine, just the delivery vehicle, in which mice are manipulated or injected in the same way as the test vaccine but without the active components.

The *positive control* can be a vaccine design that has been used in the past to protect against tumour growth.

The measurement is whether the tumour grows after delivery of the vaccine, but in these types of experiments, other control considerations may be important. For example, the actual vaccine may be composed of more than one compound, so for testing of novel vaccine delivery platforms, each component may have to be tested individually.

iii. Measuring the Efficacy of a Tumour Vaccine in People

Question: Does the vaccine inhibit tumour growth?

This question is quite different from the example above, because there are more limitations on experimental design in humans than in animals, mostly logistical, but also ethical (see Chapter 6).

Experiments in humans, designed to test therapies, go through multiple phases (Chapter 6, Figure 6.1). For this particular example, we will look at a trial where a small number (50) of cancer patients are given a tumour vaccine and we want to determine whether it works. The design is a division of the 50 patients into two equal groups – this is done without the patients or investigators knowing who is in which group. The vaccine is given to half the patients, and the other half receive no treatment. The experimental readout could be a reduction in tumour growth or number of metastases in each patient, or a long-term readout such as five-year disease-free survival. An alternative readout could be a correlation of tumour growth inhibition, such as reduction in circulating tumour DNA.

The *negative control* is the group that did not receive the vaccine. Another *negative control* could be a placebo-treated group (see Box 3.1).

There is *no positive control* for this experiment as described. It may be possible to include *a positive control*, such as a previously established vaccine in a third group of patients.

Technical Controls

In biology experiments, especially *in vitro* assays, technical controls or replicates can also be used – these are repeats of the experimental conditions within the same experiment (for example, three tubes per condition, that all get analysed separately). Because of the variability of biological systems, these are not often useful and repeats of entire experiments are a better way to control for variability. However, these technical controls can be used to ensure that the equipment being used is well calibrated (for example, pipettors) or that the person performing the experiment is consistent in their technique.

BOX 3.1 PLACEBO CONTROLS

The use of placebo controls is important in research involving interventions. The placebo effect is a scientifically recognised subconscious bias that lets the perception of a possible effect. In drug trials, participants are given the test drug and as a negative control, a placebo drug that looks the same as the test, administered the same way. This negative control is much better than giving nothing, as participants can respond to the intervention as well as to the drug.

The nocebo effect is the converse – the idea that not getting a treatment makes the outcomes worse. Both nocebo and placebo effects are considered in human trials, with the most reliable format being a double-blinded placebo controlled randomised trial. In these trials, neither the researcher nor the participant knows whether the participant is receiving a treatment or the placebo. The selection of participants into treatment or placebo groups is also done in a randomised way to prevent people using subconscious bias and accidentally putting more sick people, older people or males in one trial group or another (Figure 3.2). The placebo effect has been studied extensively, and more information is available in the Resources section.

Reporting Controls

Accurate reporting of what constitutes a control is essential for readers to determine the validity of the experiment (see Figure 3.1). If a figure is presented with error bars that represent technical controls, they are more likely to be seen as convincing and statistically significant than if they represent individual animals. Both are correct, but it must be very clear what is being represented in the data. Similarly, processed data may represent results with the negative control subtracted from each data point. Again, this is correct, but only if done accurately – presenting the negative control data is essential for the reader to interpret the data correctly.

BOX 3.2   CONTROLLING FOR TIME OF DAY IN EXPERIMENTAL DESIGN

Many experimental or even clinical trials arrange their interventions of measurements around life and other work commitments. For example, a student measuring the effect of a drug on nerve excitation in a rat model may perform a once-a-day measurement loosely every 24 hours, depending on their lecture timetable or access to the animal facility. Similarly, giving a test intervention to patients, such as a chemotherapy drug or radiotherapy regime, may not be possible at the same time every day, as it may depend on the availability of the clinical staff as well as the human volunteer. Recently, the effect of circadian rhythms and therefore time of day has emerged as an experimental variable. Both animal and human studies have shown that biological responses can differ depending on the time of day. This means that good experimental design now needs to consider controlling for time both within and between experiments. See Resources section for some specific examples.

Below are graphs of the same dataset but plotted showing variability using two different types of controls.

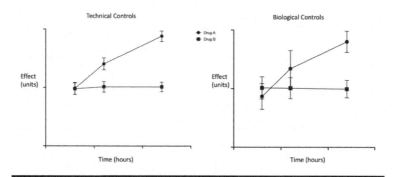

Figure 3.1   Technical versus Biological Controls.

In the dataset with technical controls, the variability in results arises from three separate tubes from one experiment. In the dataset with biological controls, the variability in results is from three different but identical experiments.

The interpretation of the effect of Drug A versus Drug B depends on which graph and therefore which controls you use.

Interpretation of technical and biological controls. Data were generated using a cancer cell line treated with Drug A or Drug B. Shown is the amount of effect on the *y*-axis and time on the *x*-axis. In the graph on the left, the same number of cells were grown in triplicate wells and treated with the same concentration of Drug A or Drug B. Shown is an average ± standard deviation of triplicate tubes at 1, 6 and 12 hours. Error bars denote variability between triplicate tubes. In the graph on the right, the same experiment was done three different times, on different days, using a different culture of the cancer cell line. The average ± standard deviation of three separate experiments are shown again as the effect over time in hours. As biological replicates compare results between individual experiments or individual mice or patient samples, the standard error increases and indicates better the variability between individual samples. How to interpret these results using statistical analyses is discussed in Chapter 4.

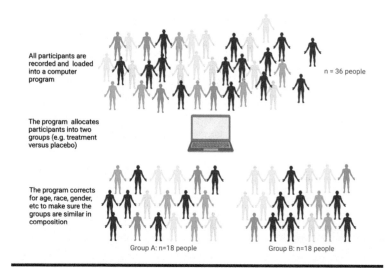

All participants are recorded and loaded into a computer program

n = 36 people

The program allocates participants into two groups (e.g. treatment versus placebo)

The program corrects for age, race, gender, etc to make sure the groups are similar in composition

Group A: n=18 people

Group B: n=18 people

Figure 3.2 Randomised Control Trials in Humans. In a randomised controlled trial, study participants are distributed into "treatment" and "control" groups randomly. This is done by randomly distributing the characteristics of patients so that each group has patients with similar characteristics. For example, details that are fed into the computer include age, gender, and stage and size of the tumour. After randomisation, each group should have about the same number of people that are female, aged 50–60 that all have a 4 mm mass (as an example). Randomisation should mitigate any biases that may influence the outcome of the study. *Figure created in BioRender.*

In human trials, randomisation of people into the test versus control group is important to reduce any bias.

Example 3.1   Flow Cytometry Controls

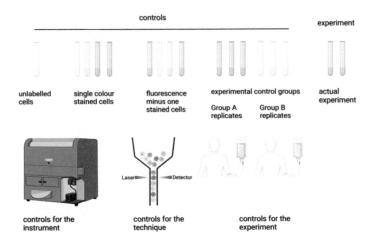

In flow cytometry, a common technique in biomedical science, antibodies are attached to fluorochromes and the antibodies bind to molecules expressed on cells, allowing the detection of different types of cells from tissues or samples. There are controls associated with antibody binding, fluorochrome spill over, the cytometer itself, as well as the normal experimental controls. There are similar types of controls for many assays used in the life sciences, so it is important to consider where in the experimental design or protocol there might be variability. *Figure created in BioRender.*

Example 3.2   Crossover Trials and What to Do When a Treatment Is Working

Randomised control trials are routinely used for testing clinical interventions, but what happens when one group seems to be benefiting from the intervention? Sometimes, researchers need to decide to allow the presumed placebo group to *crossover* to the treatment group – to withhold an effective treatment would be unethical. These decisions are never straightforward.

For example, during the development of vaccines for COVID-19, the World Health Organization published a position paper stating:

> The Working Group has concluded that although there is a scientific imperative to continue trials of vaccines against COVID-19 after a candidate vaccine is granted an EUD [emergency use designation], there is also an ethical imperative to ensure that trial participants who are at substantial risk of infection with the coronavirus SARS-CoV-2, and severe COVID-19 morbidity or mortality – such as healthcare workers at high to very high risk of acquiring and transmitting the disease, and people above 65 years of age – are in a position to access an EUD vaccine as soon as practically possible, should they wish to do so.

Singh JA, Kochhar S, Wolff J. WHO ACT-Accelerator Ethics & Governance Working Group. Placebo use and unblinding in COVID-19 vaccine trials: recommendations of a WHO Expert Working Group. Nat Med. 2021 Apr;27(4):569–570. doi: 10.1038/s41591-021-01299-5. Erratum in: Nat Med. 2021 May;27(5):925.

Potential Careers

Scientific research and teaching often requires specialised equipment and knowledge. The equipment is sold by companies who also provide training on how to use it, how to design experiments properly and how to analyse the data from the equipment. These companies often hire technical application specialists to liaise with researchers.

Technical Application Specialists

Technical specialists usually have a degree in science in an area related to the equipment involved, and often have some research experience. The role is great for people who have a technical interest in how experiments work and a commitment to helping researchers design and carry out the best research possible. The specialists often travel widely to meet clients all over the world or the region, meeting with individuals, setting up new equipment and presenting the latest capabilities at conferences. There are also application specialists who focus on software development and design. Technical application specialists remain engaged in the latest research, and so the job is perfect for someone who wants to be involved in research and who likes talking with others about their work.

Resources and Further Reading

Resources Used in This Chapter

- Clinical trials, including randomisation
  www.cancerresearchuk.org/about-cancer/find-a-
  clinical-trial/what-clinical-trials-are/randomised-
  trials
- Flow cytometry controls
  www.abcam.com/protocols/recommended-contr
  ols-for-flow-cytometry
  www.championsoncology.com/clinical-trial-soluti
  ons/clinical-flow-cytometry/high-dimensional-cus
  tom-assay-development
- Publications on ELISA and convalescent serum
  (for Exercises)

Tré-Hardy M, Wilmet A, Beukinga I, Favresse J, Dogné JM, Douxfils J, Blairon L. Analytical and clinical validation of an ELISA for specific SARS-CoV-2 IgG, IgA, and IgM antibodies. *J Med Virol.* 2021 Feb;93(2):803–811. doi: 10.1002/jmv.26303.

Joyner MJ, Bruno KA, Klassen SA, Kunze KL, Johnson PW, Lesser ER, Wiggins CC, Senefeld JW, Klompas AM, Hodge DO, Shepherd JRA, Rea RF, Whelan ER, Clayburn AJ, Spiegel MR, Baker SE, Larson KF, Ripoll JG, Andersen KJ, Buras MR, Vogt MNP, Herasevich V, Dennis JJ, Regimbal RJ, Bauer PR, Blair JE, van Buskirk CM, Winters JL, Stubbs JR, van Helmond N, Butterfield BP, Sexton MA, Diaz Soto JC, Paneth NS, Verdun NC,

Marks P, Casadevall A, Fairweather D, Carter RE, Wright RS. Safety update: COVID-19 convalescent plasma in 20,000 hospitalized patients. *Mayo Clin Proc.* 2020 Sep;95(9):1888–1897. doi: 10.1016/j.mayocp.2020.06.028.

Li L, Zhang W, Hu Y, Tong X, Zheng S, Yang J, Kong Y, Ren L, Wei Q, Mei H, Hu C, Tao C, Yang R, Wang J, Yu Y, Guo Y, Wu X, Xu Z, Zeng L, Xiong N, Chen L, Wang J, Man N, Liu Y, Xu H, Deng E, Zhang X, Li C, Wang C, Su S, Zhang L, Wang J, Wu Y, Liu Z. Effect of convalescent plasma therapy on time to clinical improvement in patients with severe and life-threatening COVID-19: A randomized clinical trial. *JAMA.* 2020 Aug 4;324(5):460–470. doi: 10.1001/jama.2020.10044. Erratum in: *JAMA.* 2020 Aug 4;324(5):519.

Gharbharan A, Jordans CCE, GeurtsvanKessel C, den Hollander JG, Karim F, Mollema FPN, Stalenhoef-Schukken JE, Dofferhoff A, Ludwig I, Koster A, Hassing RJ, Bos JC, van Pottelberge GR, Vlasveld IN, Ammerlaan HSM, van Leeuwen-Segarceanu EM, Miedema J, van der Eerden M, Schrama TJ, Papageorgiou G, Te Boekhorst P, Swaneveld FH, Mueller YM, Schreurs MWJ, van Kampen JJA, Rockx B, Okba NMA, Katsikis PD, Koopmans MPG, Haagmans BL, Rokx C, Rijnders BJA. Effects of potent neutralizing antibodies from convalescent plasma in patients hospitalized for severe SARS-CoV-2 infection. *Nat Commun.* 2021 May 27;12(1):3189. doi: 10.1038/s41467-021-23469-2.

Additional Websites That Might Be Useful

■ Overview of experimental controls
https://sciencetrends.com/experimental-control-important/
■ Technical and biological controls

Tsvetkov D, Kolpakov E, Kassmann M, Schubert R, Gollasch M. Distinguishing between biological and technical replicates in hypertension research on isolated arteries. *Front Med* (Lausanne). 2019 Jun 20;6:126. doi: 10.3389/fmed.2019.00126.

■ Crossover clinical trials
https://s4be.cochrane.org/blog/2020/09/07/crossover-trials-what-are-they-and-what-are-their-advantages-and-limitations/
■ Randomised trials
www.cancerresearchuk.org/about-cancer/find-a-clinical-trial/what-clinical-trials-are/randomised-trials
■ Placebo effect
www.testingtreatments.org/2012/09/10/the-placebo-effect/
■ NIH Policy on sex as a biological variable
https://orwh.od.nih.gov/sex-gender/nih-policy-sex-biological-variable

Research Publications for Further Reading

1. WHO Statement on placebo controls for COVID-19

Singh JA, Kochhar S, Wolff J; WHO ACT-Accelerator Ethics & Governance Working Group.

Placebo use and unblinding in COVID-19 vaccine trials: recommendations of a WHO Expert Working Group. *Nat Med.* 2021 Apr;27(4):569–570. doi: 10.1038/s41591-021-01299-5. Erratum in: *Nat Med.* 2021 May;27(5):925.

2. Crossover trials

Lim CY, In J. Considerations for crossover design in clinical study. *Korean J Anesthesiol.* 2021 Aug;74(4):293–299. doi: 10.4097/kja.21165.

3. Randomisation in clinical trials

Berger V, Bour L, Carter K. et al. A roadmap to using randomization in clinical trials. *BMC Med Res Methodol.* 2021;21:168. https://doi.org/10.1186/s12 874-021-01303-z

4. A checklist for reporting placebo and other controls

Howick J, Webster RK, Rees JL, Turner R, Macdonald H, Price A, Evers AWM, Bishop F, Collins GS, Bokelmann K, Hopewell S, Knottnerus A, Lamb S, Madigan C, Napadow V, Papanikitas AN, Hoffmann T. TIDieR-Placebo: A guide and checklist for reporting placebo and sham controls. *PLoS Med.* 2020 Sep 21;17(9):e1003294. doi: 10.1371/journal. pmed.1003294.

5. Discussion of types of placebo controls and questions about future options

Schmidli H, Häring DA, Thomas M, Cassidy A, Weber S, Bretz F. Beyond randomized clinical trials: Use of external controls. *Clin Pharmacol Ther.* 2020 Apr;107(4):806–816. doi: 10.1002/cpt.1723.

Fregni F, Imamura M, Chien HF, Lew HL, Boggio P, Kaptchuk TJ, Riberto M, Hsing WT, Battistella LR, Furlan A; International Placebo Symposium Working Group. Challenges and recommendations for placebo controls in randomized trials in physical and rehabilitation medicine: a report of the international placebo symposium working group. *Am J Phys Med Rehabil.* 2010 Feb;89(2):160–172. doi: 10.1097/PHM.0b013e3181bc0bbd.

6. Effect of time of day on experimental results

Walton JC, Walker WH 2nd, Bumgarner JR, Meléndez-Fernández OH, Liu JA, Hughes HL, Kaper AL, Nelson RJ. Circadian variation in efficacy of medications. *Clin Pharmacol Ther.* 2021 Jun;109(6):1457–1488. doi: 10.1002/cpt.2073.

Aoyama S, Shibata S. Time-of-day-dependent physiological responses to meal and exercise. *Front Nutr.* 2020 Feb 28;7:18. doi: 10.3389/fnut.2020.00018.

Exercise triggers fat breakdown at some times of day and not others. *Nature.* 2023 Feb;614(7949):596. doi: 10.1038/d41586-023-00408-3.

Pendergrast LA, Lundell LS, Ehrlich AM, Ashcroft SP, Schönke M, Basse AL, Krook A, Treebak JT, Dollet

L, Zierath JR. Time of day determines postexercise metabolism in mouse adipose tissue. *Proc Natl Acad Sci U S A.* 2023 Feb 21;120(8):e2218510120. doi: 10.1073/pnas.2218510120.

Exercises

Individuals

Exercise 1

Tré-Hardy M, Wilmet A, Beukinga I, Favresse J, Dogné JM, Douxfils J, Blairon L. Analytical and clinical validation of an ELISA for specific SARS-CoV-2 IgG, IgA, and IgM antibodies. *J Med Virol.* 2021 Feb;93(2):803–811. doi: 10.1002/jmv.26303.

This article assesses the biological assay, enzyme-linked immunosorbent assay (ELISA), for detecting anti-SARS-CoV2 antibodies. Read the paper and then answer these questions:

a. What is the difference between sensitivity and specificity?
b. What do the authors mean by "confidence interval"? Are there other tests they could have used for these data?
c. Are the patient numbers sufficient to determine an effect? How do you know?
d. Table 2 shows three levels of "interpretation criteria". Why do you think they choose these ratios?

e. Section 2.6 explains types of results that came from the data. What is "trueness"? What is "precision"? What is "carryover"? Why are these all important for this paper?

f. What is the significance of changing the cut off for: (a) patients? (b) the health system and (c) the global pandemic?

Exercise 2

Create a multichoice question that would assess a student's understanding of controls. Combine your questions from the whole class, then sit the test to see how well you do.

Class

Exercise 3

Assessing and designing a trial for plasma in COVID-19 patient

The test is whether a new treatment (convalescent plasma) can reduce the burden of SARS CoV-2 infection in hospitalised patients. First, tests are needed to confirm that the intervention is safe in a diverse population of patients and this trial is separate from the efficacy trials. Ideally, a large number of patients should be enrolled to test the safety of the intervention. In many cases, all patients receive the intervention to determine whether adverse events occur after transfusion. This may seem a bit dicey, however, in these instances, patients enrolled in these studies are not healthy subjects, but are usually suffering from the disease that would require intervention.

Once the intervention is deemed safe, then a randomised clinical trial can test efficacy. This is done by randomising patients into treatment groups and control groups; usually control groups receive standard care for the disease, or a placebo. In a recent study on convalescent plasma efficacy for COVID-19, the study was terminated due to the low number of patients enrolled. This then leads to ambiguity in the outcomes and can decrease statistical significance due to low power.

Design a clinical trial to test the new treatment.

Resources useful for this exercise:
Joyner MJ, Bruno KA, Klassen SA, Kunze KL, Johnson PW, Lesser ER, Wiggins CC, Senefeld JW, Klompas AM, Hodge DO, Shepherd JRA, Rea RF, Whelan ER, Clayburn AJ, Spiegel MR, Baker SE, Larson KF, Ripoll JG, Andersen KJ, Buras MR, Vogt MNP, Herasevich V, Dennis JJ, Regimbal RJ, Bauer PR, Blair JE, van Buskirk CM, Winters JL, Stubbs JR, van Helmond N, Butterfield BP, Sexton MA, Diaz Soto JC, Paneth NS, Verdun NC, Marks P, Casadevall A, Fairweather D, Carter RE, Wright RS. Safety update: COVID-19 convalescent plasma in 20,000 hospitalized patients. *Mayo Clin Proc.* 2020 Sep;95(9):1888–1897. doi: 10.1016/j.mayocp.2020.06.028.

Li L, Zhang W, Hu Y, Tong X, Zheng S, Yang J, Kong Y, Ren L, Wei Q, Mei H, Hu C, Tao C, Yang R, Wang J, Yu Y, Guo Y, Wu X, Xu Z, Zeng L, Xiong N, Chen

L, Wang J, Man N, Liu Y, Xu H, Deng E, Zhang X, Li C, Wang C, Su S, Zhang L, Wang J, Wu Y, Liu Z. Effect of convalescent plasma therapy on time to clinical improvement in patients with severe and life-threatening COVID-19: A randomized clinical trial. *JAMA*. 2020 Aug 4;324(5):460–470. doi: 10.1001/jama.2020.10044. Erratum in: *JAMA*. 2020 Aug 4;324(5):519.

Gharbharan A, Jordans CCE, GeurtsvanKessel C, den Hollander JG, Karim F, Mollema FPN, Stalenhoef-Schukken JE, Dofferhoff A, Ludwig I, Koster A, Hassing RJ, Bos JC, van Pottelberge GR, Vlasveld IN, Ammerlaan HSM, van Leeuwen-Segarceanu EM, Miedema J, van der Eerden M, Schrama TJ, Papageorgiou G, Te Boekhorst P, Swaneveld FH, Mueller YM, Schreurs MWJ, van Kampen JJA, Rockx B, Okba NMA, Katsikis PD, Koopmans MPG, Haagmans BL, Rokx C, Rijnders BJA. Effects of potent neutralizing antibodies from convalescent plasma in patients hospitalized for severe SARS-CoV-2 infection. *Nat Commun*. 2021 May 27;12(1):3189. doi: 10.1038/s41467-021-23469-2.

Exercise 4

Find a published journal article using laboratory experiments (as opposed to a clinical trial). From the paper, identify negative and positive, experimental and technical controls. Comment on whether the authors correctly identified the controls and used them appropriately. What other controls could you use? How would you feedback your ideas and suggestions to the authors?

Chapter 4

Statistical Analysis and Risk

Introduction and Scope

The purpose of this chapter is to discuss both the importance of statistical analyses, as well as the limitations of relying on statistical analyses to interpret data. Covering all the statistical tests and when and how to use them is beyond the scope of this chapter; these have been covered exhaustively elsewhere. However, it is vital that scientists have a basic understanding of significance, p values and how they are calculated, and to have the ability to assess whether interpretations of data are correct, based on the tests used. Finally, it is essential to consider that what is different need not be statistically significant.

DOI: 10.1201/9781003326366-4

Learning Objectives

- To consider which statistical test(s) may be appropriate to use when designing experiments
- To discuss the difference between statistically significant and biologically relevant results
- To understand p values and their use in biomedical research

When to Think about Statistics

The scope of statistical analyses is extremely broad. The purpose of this book is to help students to recognise key components of designing, performing and interpreting experiments and results. For this reason, we will only highlight some key concepts in statistical analyses, rather than summarise all facets of statistical analyses.

Designing Your Experiment

A key component of experimental design is planning the analysis strategy. Depending on the hypothesis and/or research question, ultimately the experiment is likely to compare two or more experimental parameters. The researcher needs to know in advance what they will compare and why, and to plan whether the comparison will be simple (between two similar experimental groups), complicated (multiple groups at multiple timepoints) or whether a direct comparison is the most useful way to study the data in the first place. Planning the analysis dictates subsequent experimental design, including group size, number of repeats and replicates within the experiment. A knowledge of basic statistical tests is required to work out what needs to be measured, how and why.

Analysing Results – Yours and Those of Others

Interpretation of data is the most important skill for a researcher. It is essential to understand what the experiment means, whether it is your own data or whether you are reading a published paper or watching a seminar with data. It is surprising how often researchers present statistical tests incorrectly or not at all. There is debate about best practice, but a researcher should be able to determine whether the data (i) are reliable and (ii) support the claims made. In this context, the absence of a statistically significant difference is equally valuable as the presence of one. Researchers presenting data need to first describe what they plan to consider different or not and stay consistent to that definition for the entire piece of work (for example, across different models), or justify why the definition changes (for example, a different type of experiment).

BOX 4.1   FLORENCE NIGHTINGALE AND THE ROSE DIAGRAM

Florence Nightingale is best known as the founder of modern nursing. She was also a highly regarded statistician and the first female to be inducted into the Royal Statistical Society in 1858. A master of data visualisation, Florence Nightingale became a social reformer who fought for basic sanitation practices for the British Army abroad as well as at home in Britain. Nightingale rose to prominence when she travelled to Turkey to tend to the British soldiers in the Crimean War (1854–1855). Reports from the frontlines by war correspondent William Russell described terrible living conditions and filth that was rampant in the army hospital in Scutari, part of Istanbul. With England in an uproar over the appalling conditions, the Secretary of War asked Nightingale to muster a nursing corps to travel to Turkey and help the wounded.

Nightingale was shocked by the horrible, filthy living conditions of the hospital and barracks. The hospital building was built upon a cesspool, and all manner of vermin and insects were scurrying around the hospital beds. Contaminated water, inadequate ventilation and patients lying in their own excrement added to the overwhelming burden of infectious diseases that plagued the hospital and patients. These atrocious conditions prompted the newly appointed prime minister (Henry John Temple) to assemble a Sanitation Commission comprising civilian experts that set out to improve sanitation in British army outposts and hospitals. In March 1855, improved sanitation measures and hygiene practices were incorporated at Scutari Hospital and thus the deaths attributed to infectious diseases were dramatically reduced.

After the Crimean War, Nightingale set her sights on improving living conditions for the British

Army and saving lives cut short by infectious diseases. With help from Queen Victoria, she set up the Royal Commission to investigate the health of the Army. Working with statistician William Farr, they analysed huge amounts of mortality data and realised that 16,000 out of 18,000 deaths at Scutari were due to infections made worse by poor sanitation practices! But, how best to represent the number of deaths from battle versus disease that she recorded over the time from 1854–1856?

The Rose Diagram (see figure) is also called a polar area diagram in which each circle is separated into 12 equal pieces that represent each month of the year. The circle on the right is the first year of the Crimean War, April 1854 to March 1855. Each wedge extends out from the centre and represents the number of deaths due to battle (red area), other causes (black area) or mitigable zymotic diseases or infections (blue area). The circle on the left is the second year of the Crimean War, April 1855 to March 1856 and the dotted connecting line is when the Sanitation Commission intervened. It can be seen clearly that the deaths from infectious diseases greatly outnumbered the deaths from battle and other causes. What is also quite apparent is that deaths from infection were reduced dramatically after the Sanitation Commission intervened at Scutari and provided clean water, eliminated contaminating waste, increased ventilation and furnished clean bandages and bedding.

This is an excellent lesson in data presentation and how a single graphic can clearly represent

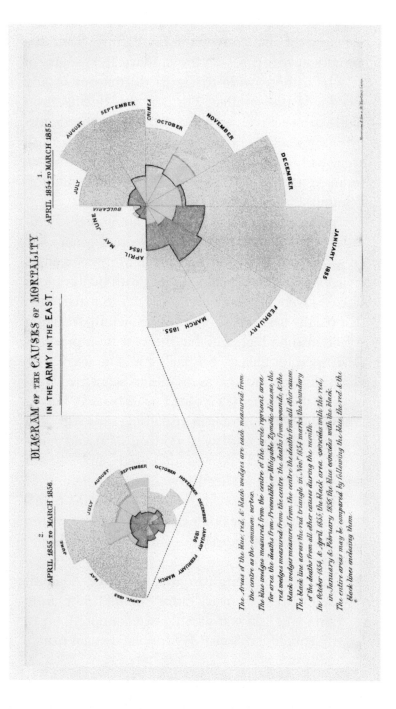

DIAGRAM OF THE CAUSES OF MORTALITY
IN THE ARMY IN THE EAST.

1.
APRIL 1854 TO MARCH 1855.

2.
APRIL 1855 TO MARCH 1856.

The Areas of the blue, red, & black wedges are each measured from
the centre as the common vertex.

The blue wedges measured from the centre of the circle represent area
for area the deaths from Preventible or Mitigable Zymotic diseases; the
red wedges measured from the centre the deaths from wounds; & the
black wedges measured from the centre the deaths from all other causes.

The black line across the red triangle in Nov.r 1854 marks the boundary
of the deaths from all other causes during the month.

In October 1854, & April 1855, the black area coincides with the red;
in January & February 1856 the blue coincides with the black.

The entire areas may be compared by following the blue, the red, & the
black lines enclosing them.

datasets yet also convey the big picture. Your eye immediately focuses on the January 1855 segment in which there were the most deaths out of any month of the conflict. Further, the deaths from battle are relatively small compared to the deaths from infection. Finally, the size of each circle (and area within each segment) clearly denotes the reduced death toll subsequent to the Sanitation Commission intervention.

The story of Florence Nightingale is one of crusader, campaigner and especially statistician. She crusaded on behalf of the British Army as well as civilians living in England where death by infectious diseases were excessive and avoidable. She reasoned that deaths from infection in the general population would also be high and that the sanitation practices developed for military hospitals could be used in civilian life. Her impact was more significant than what most of the history books lead us to believe. Importantly, she was a master in visualising data such that one graphic (picture) was worth a thousand numbers (words) and she revolutionised public health reforms to minimise deaths by infectious disease.

The graphic shows the number of deaths by infections, wounds or other causes as an area within each segment of the pie.

Image from: **Nightingale, Florence**. A contribution to the sanitary history of the British army during the late war with Russia. London: John W Parker and Sons 1859 (public domain mark).

Important Concepts in Statistics

Probability and p *Values*

The basis to statistical testing is the concept of the null hypothesis. For any given research question, for example, is a patient given Drug A responding better than a patient given Drug B?, then a hypothesis is generated: Drug A will be better than Drug B. The experimental data then provide the basis for determining whether the hypothesis is true or false. However, this is done per experiment, and *within* that experiment, a p value tells you what the probability is that your hypothesis is false. A low p value indicates a low probability that the result happened by chance, and therefore means that the hypothesis is unlikely to be false.

The size of the p value reflects the likelihood of your answer or result proving or disproving your hypothesis. Traditionally, p values of 0.05 or lower are considered to be statistically significant, although it is important to consider the research question and the type of experiment. P values are often graded, so that they are given a different category of difference 0.05, 0.01. 0.001 etc., usually indicated by stars on graphs.

Distribution

Normal distribution refers to data that fall into a familiar symmetrical bell-curve shape when plotted. This is also called a Gaussian distribution. The probability for accuracy of values at the extreme ends of the curve is lower than values in the middle. For example, height in a large population of people usually has a normal distribution. There is a higher probability that someone is 170 cm (middle of the curve) than 210 cm (at the high end of the curve). The width of the normal distribution determines the standard deviation, as measure of variability (Figure 4.1).

However, many (most) datasets collected in biological sciences, and in lab experiments, usually do not fall into a normal distribution. Because of this, inferences about variability based on median, mean and standard deviation derived from normal distribution are no longer valid. Therefore, there are different tests to measure statistical significance for normal versus non-normal distributions – parametric versus non-parametric tests. It is essential to test the dataset for normality distribution *before* a test is chosen. For example, a two-sample *t* test is appropriate for normally distributed data, but Mann–Whitney *U* test is appropriate for non-normally distributed data. The significance and testing rules are much more complicated than this simple comparison but recognising that distribution may affect the interpretation of results is key.

Power Calculations and Sample Size

In order to determine statistical significance, the groups need to have several repeats or individual readouts that are meaningful. A key determinant in differentiating meaningful data versus anecdote is the number of events that are measured. For example, if one person has no symptoms from an Epstein Barr virus (EBV) infection (sometimes called glandular fever or infectious mononucleosis), then one interpretation is that EBV does not make people sick. However, if we look at EBV infection in 100,000 young adults, we can see that 30,000 people get sick (30%). The one individual who did not get sick was one of those 70,000 infected but asymptomatic people, but that we need to include the other 30,000 who did get sick to determine an effect of infection.

Similarly, in experimental design, it is important to work out how many datapoints (for example, measurements, numbers of subjects, number of experimental repeats) will give you sufficient data to prove or disprove the hypothesis. You need to have enough data points to have a mean and a standard deviation in order to determine natural variability in the group (therefore one or two will not be enough).

Ways to Work Out the Number of Datapoints You Need

Guess

Unfortunately, this happens a lot. Scientists often default to three datapoints ($n=3$) because it is the smallest odd number that would lead to a tie-breaker decision (for example, 2/3 is a majority, therefore the data are useful). Many high-profile publications present data with $n=3$ and prove significance with statistical tests, usually making assumptions that are not accurate. Sometimes $n=3$ is right, but having a process to determine what will be enough is essential.

Educated Guess

While also not ideal, some decisions can be made based on previous similar work by the researcher or published by others. For example, if I have done a series of experiments testing Drug A with Drug B, with nine samples per group ($n=9$) and shown meaningful and accurate statistical significance, it may be reasonable to assume that an experiment testing Drug C (a derivative of Drug B) would also yield meaningful results with $n=9$. Pilot studies often rely on an educated guess as insufficient data exist to perform a more accurate determination of the number of datapoints.

Power and Sample Size Calculations

These calculations provide you with an accurate number of experimental datapoints (sample size) based on known data and a predetermined endpoint of difference (power to detect a statistically significant difference). This approach reduces the subjectivity in the experiment – the significant endpoint is already determined, and the data will either support or not support it. The endpoint can be determined from previous experiments. Power calculations are usually performed before an experiment begins.

Power calculations help to reduce statistical errors. Type 1 errors occur if the null hypothesis is rejected incorrectly – there is no difference, but the researcher reports a difference. Type II errors happen when the null hypothesis is accepted incorrectly – there is a difference, but the researcher reports no difference.

Power calculations can be affected by

(a) the variance of measurements in a sample (for example, bacteria colonies versus human patients);

(b) the magnitude of the significant difference (for example, protein concentration (pg/mL) in blood versus number of lung metastases);

(c) how important it is to avoid a Type 1 error (reporting a difference when none exists – this could lead to clinical interventions that don't work);

(d) the type of statistical test (what are the data, what are the comparisons – this is why the calculation ends to be performed first).

The sample size needs to be large enough to determine a difference, but over-sized sample sizes come with ethical, logistical and financial considerations. If ten mice have been calculated to be sufficient to determine a difference, what is the benefit of using 50 mice? In clinical studies, sample size also needs to consider the heterogeneity of a population (see Chapter 2, Figure 2.2).

Variability

Most experiments include multiple repeats of data, for example, several mice in a group or several repeats of the entire experiment. This is important to ensure reproducibility, but it also can increase the variability in the results. Variability is to be expected, but it is an important parameter to report in the results. Therefore, the best practice is to show all the datapoints on a graph rather than averaging them. Averages, confidence intervals and standard deviations can still be included, but expressing data with all the datapoints gives the reader an idea of the variability in the experiment and allows them to better draw their own conclusions.

Interpreting Statistical Tests

Discussing all statistical tests is beyond the scope of this chapter, but, for most part, biologists need to deal with two different types of data and look at the distribution of data within these two types.

The first assessment is determining a difference between two groups, A and B. This may be an experiment where one group is given a drug and one a placebo. Or it may be an *in vitro* experiment where different titrations of a molecule are given; but it generally means comparing two groups with one difference between them. The next step is to determine whether the data are normally distributed or not. The choice of statistical analysis then can be for normally or not normally distributed data.

The second type of experiment is usually comparing two or more groups, with continuous data, for example, looking at treatment between two or three groups over time. The test for normality and the number of data points are still relevant as before, but the statistical tests chosen are different because different parameters are used.

Stargazing

The value of p values in determining meaningful data have recently received attention in the scientific literature – the idea that if something is statistically significant means that it is different, is not true, and many important conclusions have been drawn from statistically significant data. In addition, many published studies have simply used the wrong test to determine significance. Now, more than ever, it is important for students to understand the tests but also to interpret the data and the test correctly.

The question is what is different? If we rely on statistical tests, then we must remember that we are still measuring the probability that something is different or not. If you have 20 results, all with a p value of 0.05, then that suggests that 1/20 experiments is not true – which one? If your data show a p value of 0.051, would you still consider that statistically significant? If the data show a p value of 0.049, would you consider that different?

There are two actions for new students to consider in this area. One is for researchers to include the actual p value on published data, rather than using a star to group p values, allowing the reader to determine whether to consider it different or not.

The second is to write consistently about differences. When reviewing a PhD thesis, it is surprising how often students talk about differences in results, but the definition of difference changes according to the narrative. If a statistically significant result is considered different, then all statistically significant results need to be considered as different. And conversely, if a result is not significant, it can't be different in one experiment and not different in another. When reviewing theses or papers, it is common to see a phrase such as "there was a trend towards more..." – this is not only mathematically incorrect, but it begins a process of bias in reporting results. When is something a trend – is it 2× more but not significant? Is it therefore a trend for everything that is 2× more? This approach to treating non-significant differences as different means that the writer has to interpret their data rather than the reader. This is simple to address by defining significant versus non-significant results. However, non-significant results certainly don't mean the result is invalid or not important!

More important is the understanding and interpretation of data and doing so with caution, and not relying on p values to determine a meaningful result. Instead, look at the research question and the limitations of the study to determine meaning.

Statistically Significant versus Biologically Relevant

Unfortunately, there are no simple tests to determine biological relevance, which is why so many scientists rely on statistical significance to interpret data. But given the variability in biological systems, especially humans, and often the limit in numbers, it is unrealistic to expect to find statistical differences in many experiments. The main point is that it doesn't matter. If there is a large difference between groups, it is fine to determine it as different and to interpret the results as such. For example, 10/10 patients had an increase in cells after treatment. Conversely, a statistically significant difference between two groups measuring physiologically low concentrations of a molecule may not be meaningful in the context of a biological system. In this situation, you would need to be cautious about determining whether the difference was meaningful or not. This is particularly common when dealing with sensitive assays. This can all be addressed by accurate reporting of results.

Risk

In biological research, particularly that discussed in the media, the term risk is used to extrapolate significant differences to determine the chance of one thing happening over another. For example, if exposure to X is statistically different to exposure to nothing, then one can calculate an individual's risk that exposure to X would lead to the same result as in the experiment.

However, there are several types of risk, and only two will be discussed here: relative risk and absolute risk. Relative risk refers to the chance that a change in something will lead to the same result; absolute risk considers the chance of this change happening in the first place. For example, if the consumption of cupcakes doubles the risk of cancer, this refers to the relative risk – you are twice as likely to get cancer. However, if the absolute risk of getting cancer is 0.00001%, then eating cupcakes doubles that risk to 0.00002% – still extremely low and not worth worrying about.

BOX 4.2   FALSE POSITIVES AND SCREENING FOR DISEASE

One of the greatest advances in modern medicine is the ability to perform testing on populations to determine early signs of disease. Examples include colorectal, prostate and breast cancer screening programmes. However, all testing, especially on large numbers of people, results in a proportion of false negatives (not detecting cancer when it is there) and false positives (detecting a cancer that actually isn't there). Because the results of screening lead to interventions, it is important to balance the likelihood of false positives with the subsequent medical advice.

For example, a 2010 study of prostate cancer screening results from Finland showed that one in eight men screened had a false positive result. This is not only stressful for the patient, but the interventions for prostate cancer can be significant, sometimes leading to incontinence or impotence. Evidence from across the world indicates that widespread screening of asymptomatic men (45–80 years old) is not recommended – for each 1,000 men screened (compared with those unscreened and diagnosed based on other means), there was one fewer death and three fewer cases of metastatic disease, balanced against no difference in overall mortality, more complications from the subsequent prostate biopsy and more complications from treatment (https://bpac.org.nz/2020/prostate.aspx).

Targeted screening of at-risk populations where possible can improve the situation – for example, targeted colorectal cancer screening of those with a family history of colorectal cancer, or those who are older – both known risk factors for colorectal cancer.

Summary

Sometimes it's fine to show data that is not statistically significantly different. Don't be afraid to show all the data and let others reach conclusions.

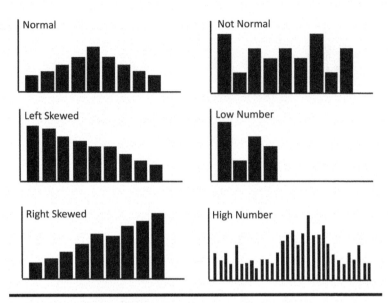

Figure 4.1   Normal Distribution.
Distribution can make a difference to data accuracy. When
planning statistical analyses of data, one of the important
considerations is the distribution of data. Put simply, normally distributed data forms a bell-shaped curve, where the
mean sits in the middle. Non-normally distributed data can
be skewed to the left or right, or just be all over the place. The
variability in non-normally distributed data mean that different
statistical tests must be used to compare between groups.
Often, small group size results in non-normally distributed
data, which is why it is important to test for the distribution
before starting to compare groups.

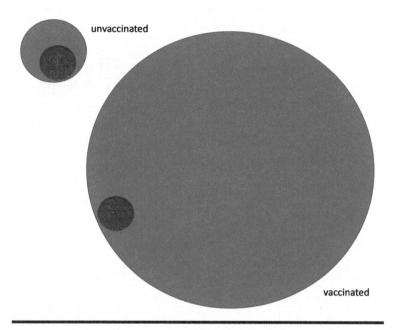

unvaccinated

vaccinated

Figure 4.2　The Base Rate Fallacy. The *base rate fallacy* is an economics term, but is relevant to many areas, including public health and even anti-terrorist policies. It refers to the analysis of numbers, for example, rates of disease, without considering the population that is being studied. This was most recently discussed in the media following COVID-19 infections and disease after vaccination. News stories and social media threads worried about seemingly high numbers of vaccinated people being hospitalised or dying from COVID-19, compared to unvaccinated people. However, these discussions did not adjust the *number* of affected people to the *rate* per category of affected people. What this means is that if you have 1,000 people in hospital who are vaccinated and 1,000 people in hospital who are unvaccinated, this doesn't mean they are the same. In a country with over 90% of the adult population vaccinated, you have so many more vaccinated people that the *percent* of vaccinated people in hospital is much lower than the percent of unvaccinated people. So, you are much less likely to be in hospital if you are vaccinated. Data from 2021 from the United States showed that across the country, you were 29 times more likely to be in hospital if you were unvaccinated than vaccinated. The picture shows the same number of people affected (for example, in hospital, red circle) is the same, but the size of the blue circle represents the population size of each group.

Example 4.1 Presenting and Explaining Data with Differences

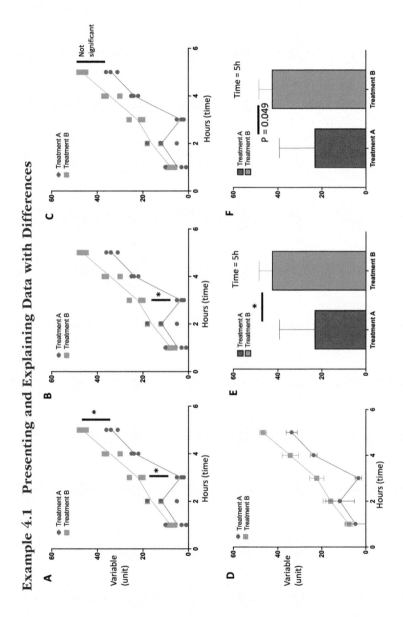

Example 4.1   Interpreting graphs.

It can be very difficult for people to talk about differences in data. If we define a difference as statistically significant, then is something that isn't statistically significant also different? Or definitely not different? What about something that seems to be increasing over time but isn't significant? Consider the graphs above, showing the same data, but with different results in terms of statistical significance. How would you talk about the results between Treatment A and B when presented with each of these graphs?

Example 4.2   The Reproducibility Crisis

The reproducibility (or replication) crisis refers to the poor translation of experimental results into meaningful outcomes. Put more simply, many experiments cannot be repeated with the same result, especially by different people, even when doing the same experiment. However, the importance in one experiment yielding a statistically significant result is often sufficient to invest more time and resources in moving the research to the next step.

There has been much discussion about the cause of the reproducibility crisis, but one strong argument is the misinterpretation of the p value as a measure of reproducibility, when it in fact "*by itself, a p value does not provide a good measure of evidence regarding a model or hypothesis*".

This last quote comes from a commentary referenced below summarising much of the discussion and potential cause of the reproducibility crisis:

> Even more important than those truths is the fact that a p-value provides no information about

whether the result should be believed, that is, that it is a repeatable finding. As a consequence, as noted in the position statement, "The widespread use of 'statistical significance' (generally interpreted as '$p \leq 0.05$') as a license for making a claim of a scientific finding (or implied truth) leads to a considerable distortion of the scientific process."

(p. 131)

So, what do we do? According to the same article, more than 90% of scientists misunderstand the p value and believe that it provides information about repeatability. A better way to demonstrate repeatability is to simply repeat the experiments.

Branch MN. The reproducibility crisis: Might the methods used frequently in behavior-analysis research help? *Perspect Behav Sci*. 2018 Jun 4;42(1):77–89. doi: 10.1007/s40614-018-0158-5.

Potential Careers

Analysis of experimental data is an essential component of scientific research – analysing your own data and being able to review data from others to determine whether it is robust or not. While it is the responsibility of the researcher to make sure experimental design is robust, and analysis is performed well, many researchers are not adequately trained to perform the correct statistical analyses for their experiments.

Biostatistician

Biostatisticians are highly specialised scientists who provide expert statistical advice to researchers performing biological experiments. This includes designing experiments and projects (for example, determining group size) as well as contributing to analysis of datasets (for example, gene expression data). Some biostatisticians specialise in clinical trials, working with researchers and doctors to design randomised controlled trials and ensuring data integrity throughout the process. Biostatisticians need to have statistics expertise but are also often required to understand both basic and specialised biology in order to work well with researchers in diverse areas.

Resources and Further Reading

Resources Used in This Chapter

■ Help with statistical testing
https://statisticsbyjim.com/hypothesis-testing/
nonparametric-parametric-tests/

Marusteri M, Bacarea V. Comparing groups for statistical differences: How to choose the right statistical test? *Biochem Med* (Zagreb). 2010;20:15–32.

■ Reproducibility crisis

Branch MN. The reproducibility crisis: Might the methods used frequently in behavior-analysis research

help? *Perspect Behav Sci.* 2018 Jun 4;42(1):77–89.
doi: 10.1007/s40614-018-0158-5. PMID: 31976422;
PMCID: PMC6701500.

■ EBV infection
  https://bpac.org.nz/bt/2012/october/glandu
  lar.aspx
■ Florence Nightingale and the Rose Diagram
  www.uh.edu/engines/epi1712.htm#:~:text=Nigh
  tingale's%20graph%20is%20like%20a,deaths%20o
  ccurred%20in%20that%20month
  https://timharford.com/2021/03/cautionary-
  tales-florence-nightingale-and-her-geeks-decl
  are-war-on-death/

From: Nightingale, Florence. *A Contribution to the
Sanitary History of the British Army During the Late
War with Russia.* London: John W Parker and Sons
1859 (public domain mark). https://wellcomecollect
ion.org/works/fecwftff

■ Base rate fallacy
  www.bbc.co.uk/programmes/p09rkn2z
  www.npr.org/2021/08/20/1029582399/planet-
  money-investigates-the-base-rate-fallacy-as-it-
  pertains-to-the-pandemic
  https://en.wikipedia.org/wiki/Base_rate_fallacy
■ Reproducibility crisis

Branch MN. The reproducibility crisis: Might the
methods used frequently in behavior-analysis research
help? *Perspect Behav Sci.* 2018 Jun 4;42(1):77–89.

doi: 10.1007/s40614-018-0158-5. PMID: 31976422; PMCID: PMC6701500.

■ Screening for cancer
https://bpac.org.nz/2020/prostate.aspx

Additional Websites That Might Be Useful

■ A book describing the impact of poor choices in experimental design and analysis

Rigor Mortis: How Sloppy Science Creates Worthless Cures, Crushes Hope, and Wastes Billions, www.goodreads.com/en/book/show/34799271-rigor-mortis

■ A podcast discussing whether statistical testing should be used
www.abc.net.au/radionational/programs/healthreport/getting-rid-of-statistical-significance/11470502
■ A summary of common statistical mistakes and how to avoid them

Riley RD, Cole TJ, Deeks J, Kirkham JJ, Morris J, Perera R, Wade A, Collins GS. On the 12th day of Christmas, a statistician sent to me… . *BMJ.* 2022 Dec 20;379:e072883. doi: 10.1136/bmj-2022-072883. PMID: 36593578; PMCID: PMC9844255.

■ How to choose the right statistical test
www.biostathandbook.com/testchoice.html

■ Base rate fallacy
www.npr.org/2021/08/20/1029582399/planet-
money-investigates-the-base-rate-fallacy-as-it-
pertains-to-the-pandemic

Research Publications for Further Reading

1. A commentary examining the rigour of research findings

Ioannidis JPA. Why most published research findings are false. *PLoS Med.* 2005;2(8):e124. https://doi.org/10.1371/journal.pmed.0020124

2. Choosing the right tests

Jones SR, Carley S, Harrison M. An introduction to power and sample size estimation. *Emergency Med J.* 2003;20:453–458.

3. Normal distribution

Mishra P, Pandey CM, Singh U, Gupta A, Sahu C, Keshri A. Descriptive statistics and normality tests for statistical data. *Ann Card Anaesth.* 2019 Jan-Mar;22(1):67–72. doi: 10.4103/aca.ACA_157_18. PMID: 30648682; PMCID: PMC6350423.

4. Cancer screening

Tikkinen KAO, Dahm P, Lytvyn L, Heen AF, Vernooij RWM, Siemieniuk RAC et al. Prostate cancer screening with prostate-specific antigen (PSA) test: a clinical practice guideline. *BMJ.* 2018;362:k3581 doi:10.1136/bmj.k3581.

Kilpeläinen, T, Tammela, T, Määttänen, L, et al. False-positive screening results in the Finnish prostate cancer screening trial. *Br J Cancer* 2010;102:469–474. https://doi.org/10.1038/sj.bjc.6605512

Hoffman RM. Implications of the new USPSTF prostate cancer screening recommendation – attaining equipoise. *JAMA Intern Med.* 2018;178(7):889–891. doi:10.1001/jamainternmed.2018.1982.

Exercises

Individual

Exercise 1

Here are a series of graphs and tables showing the same data set. Data is represented in a table format (A) or graph format (B, C). The data show the frequency (percent of total cells) of a cell type and the expression level (MFI) of a protein on those cells.

Discuss the advantages and disadvantages of displaying data in each format. What are the strengths and limitations of data in a table vs a bar graph? Does a simple bar graph (B) convey all the information needed to interpret these findings? Are the error bars in B the same as the error bars in C? What might account for this difference? In other words, how should the error between samples be calculated?

A.

Group	% CD11c/Class II^{hi}	CD86 MFI
media	4.78 %	4365
	8.00 %	4363
	3.73 %	4448
Average	5.503	4392
treatment A	6.57 %	8260
	6.49 %	8888
	1.49 %	7325
Average	4.850	8158
treatment B	6.53 %	8847
	6.00 %	8911
	1.83 %	7669
Average	4.787	8476

B.

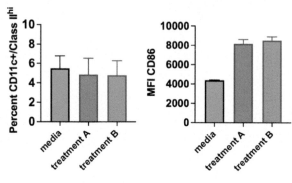

C.

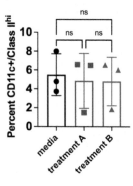
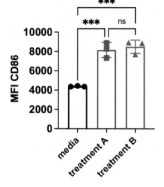

Exercise 2

You have designed an experiment to measure tumour size in a mouse model. You are comparing tumour size in a group of mice treated with saline versus a group of mice treated with a new drug. You did a pilot experiment, with $n=6$ in each group and measured tumour growth at day 10 after treatment. In the pilot experiment, you found that the control group had a mean tumour size of 100 mm², with a standard deviation of 13 mm². In the treated group, you found that the mean tumour size was 80 mm² (20% difference). You can assume the type I/II error rate is 5%. You can also assume that the ability to determine a difference if one exists (power) is 80%. Calculate the sample size you need to perform the experiment.

https://clincalc.com/stats/samplesize.aspx

Now, try different variables and scenarios. Change the group sizes. Change the power. What if you measure whether they get a tumour or not, rather than tumour size?

Class

Exercise 3

A pharmaceutical company is trying to test their new drug developed to fight diabetes. Their drug A is being tested head-to-head with a widely available drug currently on the market (drug B). In a small clinical trial, glucose levels were measured daily in a cohort of 20 people given drug A and in 20 people given drug B. The table below shows the recorded data:

Cohort 1 – Drug A	Blood glucose levels (mg/dL)	Cohort 2 – Drug B	Blood glucose levels (mg/dL)
Subject 1	120	Subject 21	71
Subject 2	81	Subject 22	67
Subject 3	101	Subject 23	69
Subject 4	79	Subject 24	90
Subject 5	58	Subject 25	61
Subject 6	118	Subject 26	103
Subject 7	64	Subject 27	67
Subject 8	52	Subject 28	91
Subject 9	79	Subject 29	61
Subject 10	103	Subject 30	85
Subject 11	79	Subject 31	88
Subject 12	85	Subject 32	109
Subject 13	56	Subject 33	62
Subject 14	72	Subject 34	68
Subject 15	62	Subject 35	81
Subject 16	72	Subject 36	96
Subject 17	118	Subject 37	101
Subject 18	66	Subject 38	95
Subject 19	105	Subject 39	96
Subject 20	110	Subject 40	105

Describe how you would graph these data and how you might test for statistical significance.

Exercise 4

Choose a research paper each and identify a figure in the paper that shows data with statistical tests applied. Why do you think they used these tests? Are they accurate? What concepts should have been considered by the authors. Now, rate the statistical analyses in the paper from 1 (low) to 10 (high). Rank the class's papers from low to high and discuss what makes good versus bad statistical analysis. Discuss how the statistical analysis contributed to the conclusions for each paper.

Chapter 5

Communicating Your Science

Introduction and Scope

The purpose of this chapter is to introduce you to the different ways that scientists use to communicate their research findings. Over the course of your scientific career, you will be asked to share your research in a variety of forms, in seminars and conference symposia, or data shared in posters, manuscripts and theses. In this chapter, we discuss the many parameters that need to be considered when presenting data, including supplementary data, representative samples, manipulated data and the dangers inherent in the decision-making process. We also highlight global reporting standards to guide these decisions. Finally, we provide some best practice for data presentation in different media.

DOI: 10.1201/9781003326366-5

Learning Objectives

- To critically assess the efficacy of different representations of data in communicating results
- To compare and contrast the guidelines regarding data for an oral or poster presentation versus a research manuscript
- To define and discuss the importance of reporting standards on the rigour and reproducibility of results

Data Presentation

In order to communicate scientific studies, the presentation of the data is key. Graphs and tables need to be easily interpretable by the audience and have to clearly and correctly display the data. Given that new technologies are not always suitable for traditional paper publications, and that big data is not feasible to display in conventional graph format, there is renewed research in other media, such as videos, code and large public datasets. However, the variability inherent in these approaches means that standardisation and guidelines for publication have become increasingly important.

Presenting Data in Different Formats

Presenting Data in a Poster Format

For most research students, the first official presentation of their data usually happens through a poster at a scientific conference. While it is equally important to tell a narrative as in an oral presentation, the scope for including data is much more limited than in a poster. The how-tos for creating an effective poster are shown in Box 5.1.

Posters are informal publications and are not held to the same standards as a normal paper publication. They are usually used to present research in progress, and so an incomplete story is acceptable. Generally, posters are presented in an informal setting with the researcher present to explain the rationale and details of the study and to answer (and ask) questions.

However, the short format of a poster means that selection of data is important – you need to choose which data you will present and which you will exclude. Too much data is impossible to display clearly, and presentations are all about making it easy for the audience to get excited, and engage with you, about your work.

BOX 5.1   TIPS FOR AN EFFECTIVE POSTER PRESENTATION

Tip 1 – Appropriate size and scale. Posters are usually presented in A0 size but can be landscape or portrait (check if there are rules on this for each presentation forum!). Make all the fonts big and graphs easily discernible from a few steps away. An easy way to check is to prepare your poster as a single slide in PowerPoint, then zoom to 400% – this will be real life A0 size.

Tip 2 – Good formatting and inclusion of diagrams are important. Graphics make it easier for the audience to understand difficult or complex ideas, such as a method flow chart or a summary diagram. Don't try to make your background a coloured picture, it's too distracting at that size. Use colours to outline boxes so people can follow the flow of information.

Tip 3 – Use of colour. Coloured graphs are more visually appealing than black and white for posters. However, stick with your colour theme throughout the poster – if your control sample is a black bar or symbols, treatment 1 is a blue bar and treatment 2 is a red bar, keep that same colour scheme in every graph.

Tip 4 – Avoid repeating information. Some presenters use the poster to state the abstract, however, if the conference or proceedings has an abstract book, then that space can be better used for an introduction or to set up the objective of the work.

Tip 5 – Talking to your poster. When presenting your poster to an audience, remember they are passing through and are on a time limit. Have a five- to ten-minute overview prepared and ask participants if they would like you to go through the poster. Then, hit the highlights without getting bogged down in too many details.
See Figure 5.1.

Presenting Data in an Oral Format

Data is usually first discussed in an informal or local scientific meeting, such as a lab meeting or departmental seminar. The how-tos of creating an effective presentation are shown in Box 5.2. An oral presentation is about making sure people follow you from the beginning to the end. Data need to be presented clearly, without too much on one slide and with helpful headings, and labelled graphs. However, the first step is to create a narrative so that an audience can follow the story you are trying to explain.

The first rule of a presentation is that nobody remembers anything you have presented in the past, so you need to set the rationale and background, as well as experimental design before you start with results and data. This is true even in a small meeting. Think of the person sitting next to you and the last time you saw them present. Do you remember the statistics of the disease they study? Do you remember all of the steps of the signalling pathway they look at? Do you remember how many mice they put in each group?

It is also important to put each presentation in context – who are you speaking to. Is your audience diverse with different backgrounds in science? Are you presenting for a class, discussing your summer research or presenting the work of others in a Journal Club format?

For each presentation type, there are different guidelines that should be used to convey your message in a concise, clear and visually appealing way. No one wants to sit through a seminar in which three or four figures are on each slide, the background colour blends in with the text colour and people in the back of the room cannot read the graph axes or titles. In the age of PowerPoint and other graphics programs, there is tendency to be fancy, adding coloured backgrounds, animations that zip graphs and symbols into the frame and cramming too much information on each slide.

Although these first presentations tend to be informal, taking time to create effective presentations with a complete narrative (background/rationale, experimental design, results and interpretations, and conclusions and future studies) provides you with experience and positions you well for more formal presentations in the future.

**BOX 5.2   TIPS FOR AN EFFECTIVE
ORAL PRESENTATION**

Tip 1: Figure out your message and stick to it. Depending on the length of your talk, you want to give the audience key pieces of information – the take home message – while also giving enough background and methodology for the audience to understand your message. Part of this rule is knowing your audience (undergraduate peers, research project mentors, faculty, lay persons, scientists or a mix of all these) to ensure that your talk can be understood by a majority of people in the room.

Tip 2: Craft your slides for your narrative. The slide should contain only the information that you talk about. Anything else is distracting from your message and confusing to the audience. If you have a nice background slide on a signalling pathway, but you only talk about three proteins or kinases in the pathway, then block out the other proteins so those images don't detract from your summary of the slide.

Tip 3: Each slide should have one conclusion and that conclusion should be the title of the slide. It's very easy to use descriptive titles to your slides such as "Western Blot Analysis", but that tells the audience nothing about the results of that analysis. Instead, lead the audience to your conclusion by having the title of the slide say something like: "Western blot analysis shows an increase in protein [X] with treatment [Y] over control treated cells". This strategy should also be used when doing presentations such as Journal Clubs and when you are presenting a paper for class.

Tip 4: Make slides easy to see and read. Use fonts that are sans serif (Arial, Helvetica, Calibri) because they tend to look better from farther away. Don't use too many different colours or fonts of writing, and use italics or one other colour to allow text to stand apart. Be aware that underlining and italics can make the text difficult for dyslexic people to read (see Example 5.1).

Tip 5: Usually, use one piece of data or graph per slide and no more than two graphs on a slide to make your point. You want to give the audience time to digest and consider your data. Most of the time, one graph is enough to show that conclusion. For presenting other work, such as Journal Club, *do not* try to cram the whole figure on one slide.

Tip 6: Explain your data. Every graph needs an orientation – what is the x-axis? What is the y-axis? What are the groups? What are the comparisons to be made? Remember this is the first time the audience has seen this graph and they need to know how to read it.

Tip 7: Repeat your message throughout the talk. Structure your talk so that you can chop it up into different sections, each with its own introduction and conclusions. Or use an outline of the talk (for example, a list of points) as a roadmap that shows the audience what might be coming next.

Tip 8: Stay on time. Prepare the talk for the time that is allotted, usually reserving some time for questions at the end.

Tip 9: Animation isn't always useful. An animated presenter is better than an animated presentation.

Tip 10: A picture is worth a thousand words.
This is particularly useful for complicated methodologies – a flow chart or diagram can be very
effective. There is time in the questions for the
audience to ask for specific details.

Presenting Data in a Paper

The ultimate goal for disseminating research
findings is to publish it in a peer-reviewed scientific
journal. This format gives you more room than
a poster, but less freedom than a talk. Again, the
narrative is key and arranging your data in the
order that best communicates a story is much more
effective than arranging it in the order that you did
it. Some students get concerned about changing
data narratives but remember that it is about
communicating data – changing the data itself is
terrible; rearranging the order in which you present
it is helpful. Most journals provide details on how
much data and what styles they prefer for their
submissions.

As for the oral and poster presentations, data
need to be clearly and accurately presented. Small
figures or figures where the legends or formats are
inconsistent confuse the reader and may obfuscate the
data. As discussed in Chapter 4, it is best practice to
show all data points (variability) rather than averaging
data. There are several formats of graphs that allow
you to do this.

Data can be presented in multiple ways. The reasons to choose a type of graph or table are beyond the scope of this book. However, because humans naturally recognise patterns, then a graph is usually more helpful for comparing parameters than a table; however, a table is sometimes needed for complex lists of groups and comparisons.

The type of graph you choose should best reflect the result you are trying to show, without hiding or minimising variable or weak results. Comparisons of groups should be presented as different colours or shades, but they need to be very distinct (avoiding green versus red for colour-blind people). Similarly, dot plots should use large symbols, easily distinguished from each other, for example, black square versus white circle, not triangle versus diamond. Journals have strict rules about the size, format and types of figures they allow, so researchers must check these for each journal before submission. Because journals usually charge money to publish coloured figures, creating effective black and white graphs is a good skill to learn. As you analyse experiments, design a graph template that works in black and white as well as colour, and use this at all stages of analysis and presentation, rather than creating brand new graphs for a thesis or paper.

Presenting Data in a Thesis or Dissertation

Generally, the advice for presenting data in a publication should be followed for a thesis or dissertation. The exception is that a thesis has much more space than a publication. While figure quality and data narrative are important, there is also scope to spread out data over multiple figures. This allows the reader to follow the narrative more easily. Typically, you also have the space to include development and optimisation of methods and analyses.

Choosing Data to Present

Representative Data or All the Data?

Biological experiments typically involve multiple repeats of the same experiment. This approach is necessary to ensure scientific rigour but also means that the amount of raw data can be massive. How do you present the data? There are effectively three options:

Show Representative Data

This means that you choose the experiment or set of results that is typical of the whole set. For example, if you designed an experiment to compare an intervention in two groups of mice and did this experiment five times, you would present the results of the two groups of mice in one of the five experiments, assuming that all five gave similar results.

Show All the Data

This means that all of the experiments are shown. This can be quite cumbersome but is often necessary if there is large variability between experiments or samples; this often happens with clinical research involving individual patients. Data can be summarised to show an average with individual points; experiments can be pooled onto one graph (as long as experimental variation can be seen in the graph), or multiple experiments/patients can be shown as individual graphs. Many journals will not accept experiments where individual data points are not shown.

Show Some of the Data and Put the Rest in Supplementary Materials

The data that are essential to tell the story of the result are included in the main body of the paper, but supporting data, such as the other four experiments, can be included in supplementary data. Supplementary figures are not included in the published paper but are linked to it and therefore available for every reader to access. This is a useful way of showing technical data in how you performed your experiments, such as flow cytometry gates, or raw PCR data. Some journals have limitations on the amount of supplementary data. Many journals require the authors to make all their raw data available as well – this ensures scientific rigour can be established by readers; for example, making every microarray data file available. There are several free online depositories that facilitate this.

In the past, publishing restrictions have led to the absence of negative results in the literature.

Fortunately, with digital online repositories and space for supplementary data, it is now much easier to include all the data in the study, including those that did not lead to a novel and exciting result. These negative data are essential for readers to put the work into context and are important because a no-result can be as informative as a positive result. If a treatment doesn't change anything, then it's useful for others to know to (i) not use that treatment and (ii) not repeat the exact experiments you did to get a similar negative result, which can be financially and ethically costly.

Global Reporting Standards

Given the increased use of technology in science, it becomes harder for individual reviewers to maintain expertise in the technical aspects of all experiments. We can see papers published using a technique we are familiar with but done to a lower standard that what we would have done. Many journals prefer to use global reporting standards to ensure that all the required information is available to determine the quality of the data, either for the reviewer, or for readers after publication. An example of this is flow cytometry guidelines (MIFlowCyt: the minimum information about a flow cytometry experiment. Cytometry A 2008 Oct;73(10):926–30.doi: 10.1002/cyto.a.20623) that includes a checklist of technical details for flow cytometry experiments.

Data Manipulation

The availability of new graphics and programmes gives researchers more freedom to design their data presentation; however, there is also an increasing problem with data manipulation. This may include cosmetic manipulations such as darkening or lightening Western blot bands; cut and paste of flow cytometry dots, or data massaging, such as removing an outlier point from a group. Many of these manipulations can be justified by the researcher, but they are never okay. Many journals now have explicitly stated image integrity standards that all authors must follow when submitting a manuscript for publication. This ensures that all submissions follow a common procedure for handling image data (www.nature.com/nature-portfolio/editorial-policies/image-integrity).

If you have five representative experiments and one clearly shows an excellent result, except that it has an outlier, a researcher may be tempted to remove the outlier, based on their knowledge of the other four experimental repeats. The drive to show statistical significance is often the rationale (see Chapter 4); however, simply showing all the data and letting the reviewer or reader determine their own significance is a much better approach.

Title that explains the exciting topic and results

Authors – it's good to highlight the person presenting
Contact information is smaller

Introduction

- Why is your research important? What problem are you trying to solve?
- What is known already about this problem?
- What new approach or angle are you investigating?
- A picture can be very helpful

Hypothesis and Aims

Many people will shortcut straight to here so be clear about what specifically you are testing

Methodology

- Flowcharts and diagrams are most useful
- If people want specific details, they will ask

Result 1 – State the result in the title, not the method

- Make sure axis labels are easy to read
- Make sure graphs are clear, well resolved and that groups can be easily identified
- Make the legend concise

Result 2

Numbering results boxes makes the poster easier to navigate

Result 3 – boxes can be different sizes

Summary

- What are your main findings?
- Do they address the hypothesis and aims?
- A diagram of a model can be useful

Conclusions

Conclusions are what you think, not what you did

- Why are your data cool?
- Why is your result important?

Figure 5.1   Example Poster. A poster is about conveying simple information and engaging the audience. Posters are almost always presented in a format where the scientist can talk to people while pointing at the poster. This means that lots of details, especially in methodology, can be left out. If people want to know, they will ask! At the same time, your poster needs to be able to tell a story without you there – titles, especially for results sections, are important and make sure the titles are meaningful. Can a person work out your whole story from just reading the section titles? Figure created partly in BioRender.

Example 5.1   Designing Inclusive Slides

More than 10% of the population has dyslexia or other issues with reading and learning. This represents a significant part of any audience, and so should be considered when creating slides to convey information. Simple ideas for improving the text on slides to make them accessible to everybody include:

- Avoid red/green contrast for those with colour blindness.
- Use left justification for text and try not to make sentences too long – Microsoft recommend the maximum number of text lines is 7 and the maximum number of words per line is 6.
- Make sure the slide is organised in a clear way, with logical progressions of ideas that are neatly arranged.
- Use a simple font, dark colour on a light background, but avoid a bright white background.
- Keep the colour consistent for each block of text.
- Underlining and italics can be hard to read as the letters tend to run together.

BAD PRESENTATION
BINGO SM

Text-heavy slides	Confusing graphics, charts	Zips thru too many slides	Use of jargon	Reads slides out loud
Facing screen, not audience	Introduction of introducers	Reads a written talk	Uses Laser Pointer	Glued to podium, stiff as a corpse
Struggles with technology	Excessive data	FREE	Runs long; no time for Q&A	Long tangents
Disorganized rambling	No eye contact with audience	Cheesy PowerPoint graphics/ templates	Starting late	Talking at slides with the pointer
No plot, characters or storyline	Lacks enthusiasm	Speaks too softly; no mic	Monotone voice	Small fonts (<20pt)

Bad Presentation Bingo is inspired by the Illinois Science Council to encourage presenters to be considerate of their audiences (especially public audiences) by paying attention to a presentation's format and delivery as much as to its content.

Figure 5.2 Presentation Bingo. Presentation Bingo is a fun way to assess an oral presentation. You would be surprised how many faculty members as well as students, sometime secretly do this in meetings or conferences. However, Presentation Bingo is a really useful peer-assessment tool when a group of students must present their data, for example, as an assessment or for a symposium or conference. When someone is practising the talk, the audience can feed back on things that got flagged during Presentation Bingo, for example, too much text, or excessive data. Figure used with permission.

Example 5.2 Requirements for Journal Submissions

Because journal articles can cover a range of topics, experimental protocols and analyses, it can be difficult for peer reviewers to determine the quality of the data and accuracy of the conclusions. Some journals now include a submission checklist that authors must complete. The checklist collects all the background data for the experiments shown. This requires authors to list their statistical tests, the number of samples or animals, clinical trial ethics, the details of the equipment used, etc. This means that the article submission can be independently assessed for statistical quality before it reaches peer reviewers. Then the reviewers can concentrate on the science, knowing that the article has passed the basic requirements of quality. For example, see the *Nature* reporting checklist for life sciences articles.

Potential Careers

Communicating science to funders, other scientists and the public is often difficult. Using images, diagrams and infographics can help to simplify a complicated scientific result or story.

Science Illustrator

Science illustrators create images and graphics for journal articles (for example, graphical abstracts), presentations or textbooks. They are also used for teaching students. Science illustrators have expertise in graphics programs and art, but often also have knowledge of scientific principles, so they can accurately portray a scientific concept. An additional aspect of science illustration is software development, creating programs that allow researchers to easily create their own diagrams, such as BioRender. These jobs are great for those who enjoy the creativity of science and enjoy communicating with others.

Resources and Further Reading

Resources Used in This Chapter

- Presentation Bingo
  www.monicametzler.com/bad-presentat
  ion-bingo/
- Image integrity for publications

www.nature.com/nature-portfolio/editorial-polic
ies/image-integrity
- Considering dyslexic audiences for slide
  presentations
  www.brightcarbon.com/blog/powerpoint-presen
  tations-and-dyslexia/
- Science illustration
  www.biorender.com

Additional Websites That Might Be Useful

- How to make a good slide presentation
  www.urmc.rochester.edu/MediaLibraries/
  URMCMedia/labs/frelinger-lab/documents/Presen
  tationFrelinger.pdf
- How to give a good oral presentation
  https://biology.indiana.edu/student-portal/semin
  ars/how-to-give-a-seminar.html
  www.youtube.com/watch?v=SFxVihJ1KSo
- How to make a good poster
  https://guides.nyu.edu/posters
- How to make good graphs
  https://scc.ms.unimelb.edu.au/resources/data-
  visualisation-and-exploration/data-visualisation
- Journal submission checklist – everything you
  need to think about when preparing a manuscript
  for publication
  www.elsevier.com/__data/promis_misc/Submiss
  ion%20Checklistdraft%20Final%20ARRCT.pdf

Research Publications for Further Reading

1. Flow cytometry standards

Lee JA, Spidlen J, Boyce K, Cai J, Crosbie N, Dalphin M, Furlong J, Gasparetto M, Goldberg M, Goralczyk EM, Hyun B, Jansen K, Kollmann T, Kong M, Leif R, McWeeney S, Moloshok TD, Moore W, Nolan G, Nolan J, Nikolich-Zugich J, Parrish D, Purcell B, Qian Y, Selvaraj B, Smith C, Tchuvatkina O, Wertheimer A, Wilkinson P, Wilson C, Wood J, Zigon R; International Society for Advancement of Cytometry Data Standards Task Force; Scheuermann RH, Brinkman RR. MIFlowCyt: the minimum information about a flow cytometry experiment. *Cytometry A*. 2008 Oct;73(10):926–30. doi: 10.1002/cyto.a.20623.

2. The effects of poor reporting in scientific publishing

Kilkenny C, Parsons N, Kadyszewski E, Festing MF, Cuthill IC, Fry D, Hutton J, Altman DG. Survey of the quality of experimental design, statistical analysis and reporting of research using animals. *PLoS One*. 2009 Nov 30;4(11):e7824. doi: 10.1371/journal. pone.0007824.

Vesterinen HM, Egan K, Deister A, Schlattmann P, Macleod MR, Dirnagl U. Systematic survey of the design, statistical analysis, and reporting of studies published in the 2008 volume of the Journal of Cerebral Blood Flow and Metabolism. *J Cereb Blood*

Flow Metab. 2011 Apr;31(4):1064–72. doi: 10.1038/jcbfm.2010.217.

Exercises

Individual

Exercise 1

Find a poster online or from your department or institution. What do you like and what don't you like about it? Is it easy to read? Is the message clear? What feedback would you provide to the author to improve the poster?

Exercise 2

Visit BioRender.com, one of the best resources for creating scientific illustrations. What images or icons do you think they could include? Prepare an argument for BioRender to support inclusion of that icon into their catalogue.

   a. Icon name
   b. List the main identifying features that must be shown in this icon
   c. Any other details

Class

Exercise 3

Many journals now provide a graphical abstract of the research. This is in addition to the conventional written abstract summarising the work. Choose a research article and prepare a graphical abstract – this can highlight the methods used and/ or the conclusions drawn from the research. For examples, see the latest publications from *Cell* (www. cell.com/cell/current)

Exercise 4

Using the data from Chapter 4, Exercise 4.3, create a series of graphs in different formats. Decide which would be best for a poster versus an oral presentation. Why? Why not? What other factors could you consider?

Chapter 6

Ethics in Science

Introduction and Scope

The purpose of this chapter is to introduce you to
the responsible conduct of research and highlight
why it's important for scientists to think about
the moral and ethical implications of their work.
Thinking responsibly, avoiding conflicts of interest
and considering various institutional policies are all
designed to help you conduct experiments in ways
that can be interpreted correctly and avoid most
ethical dilemmas. This chapter introduces you to the
three key areas of ethics in science: the use of animals
in experiments, the use of humans in experiments and
the regulation of ethical guidelines.

 DOI: 10.1201/9781003326366-6

Learning Objectives

- To understand the ethical considerations of performing research with animals
- To understand the ethical considerations of performing research with humans and/or clinical samples

Why Scientists Need to Think about Ethics

Scientists, both researchers and teachers, are faced with ethical decisions daily. While an individual's moral standard may differ from that of other people, the ethical approach to science and scientific reporting is held to a unified standard, recognised and legally regulated. Research using animals or human subjects are obvious examples of the need for ethical standards and monitoring, but other ethical issues are less clear. When is it appropriate to declare a conflict of interest? How does one justify an experiment causing pain or distress? How are different cultural perspectives handled? Is it possible to withhold a known treatment to test a new treatment? Is it unethical or immoral or illegal?

Animal Research

The use of animals in research has proven essential in making progress in science. Animals represent an entire physiological system, which cannot be completely replicated with *in vitro* culture systems or *in silico* computer modelling. Animals that are used in science include simple organisms such as zebrafish, to more complex organisms like frogs, fish, octopus, mice, rats and ultimately non-human primates such as monkeys. Regardless of complexity, the use of any animal of any species for research should be considered a privilege, not a right, and every individual animal needs to be treated with respect and care.

Three Rs of Animals Testing

The three Rs refer to refinement, reduction and replacement. When animals are considered for research purposes, the investigator must determine whether there is a *replacement* for animals, such as tissue culture or *in silico* modelling; *refinement* – ways to make the experiment more efficient; and *reduction* – ways to reduce the number of animals needed but retain the ability to make sound conclusions from the work. For example, a study using a small number of animals may not give significant results, therefore the animals have not been used properly; but over-estimating the number of animals needed is also unethical (see Chapter 4). Determining the number of animals required should be a first step in experimentation and not be merely a guess. Further, experiments must be performed and reported well – there is a major ethical problem with scientists repeating the work of others and using animals to do so when the results have been published before. Again, this can be difficult – science relies on external validation, but simple repetitive experiments may not be ethical when using animals.

There is often criticism of the use of mouse models in science. Mouse or rat models are used as the standard for analysing outputs for a disease, but often these models are used partly because they serve as validation of previously published work, and partly because techniques and reagents have been optimised. However, many models do not accurately reflect the disease process or the biology of humans. Designing, validating and sharing a new animal model is time-consuming, expensive and comes with its own ethical cost in the testing.

Non-Rodent Models

A major ethical issue with scientific experiments on animals is the psychological experiences of the animal. Rodents are often used because of their size, convenience and cost. Correct animal husbandry includes enrichment procedures to ensure that the quality of life of animals is minimally affected, for example, providing a dark hiding place for mice to retreat to and feel safe. However, animals like primates and even octopuses are also used because they are a more accurate reflection of humans. These animals have much greater sentience than rodents. Not only are primate studies difficult from a practical and financial perspective, the psychological effects on the animal in captivity must be considered and justified. Fortunately, there is a wealth of veterinary information for scientists to access to determine how they can reduce suffering and stress of experimental animals, ultimately resulting in more physiologically accurate experimental results.

Recording Pain or Distress

It can be difficult to determine endpoints in animal experiments. In many countries, death as an endpoint is not permitted, nor is extreme suffering. From a scientific perspective, these restrictions can limit the interpretation of experimental results, but it is important to design experiments that still allow you to determine a result and ensure the animals don't suffer. For example, in cancer research, mice are often given a tumour cell line and the tumour grows in the mouse. The mice must be euthanised before the tumour gets so large that the animals are in pain or impaired in their normal life. To measure the anti-tumour effect of a drug, you may have to choose a tumour size that isn't at the same point as you would be treating a human. In these instances, correlates of protection, for example, measuring anti-viral antibodies in the blood, could be used to determine the effect of the drug without allowing the tumour to grow to an unethical size.

BOX 6.1   AN EXPERIMENT ON A BIRD IN THE AIR PUMP

The famous painting "An Experiment on a Bird in the Air Pump" was painted in 1768 by Joseph Wright. It captures an experiment using a vacuum pump to measure the importance of air for living animals. Robert Boyle was a chemist in the 17th century who designed the vacuum pump to measure the properties of air. He used the pump for a variety of experiments – measuring sound, combustion and magnetism. Boyle is considered to have developed the Scientific Method (see Chapter 2) to address research questions. The use of animals in experimentation was commonplace at this time, with little consideration of the ethical questions about experimenting on living, sentient beings. Interestingly, even at the time, some people watching these experiments were uncomfortable with the use of animals, and many scientists who performed these experiments for fee-paying audiences used a bladder filled with air instead.

Image sourced from: https://en.wikipedia.org/ wiki/An_Experiment_on_a_Bird_in_ the_Air_Pump

Generation of Animals for Research Purposes

It must be remembered that while animals are used as tools for research, they are in fact alive and have a right to be treated well. Recent advances in technologies means that large-scale genetic mutation experiments can be applied to animals such as drosophila, zebrafish and even mice. These approaches can involve breeding multiple mice with different mutations to see what you get and then analyse the resulting phenotype generated from a random gene mutation. While there is significant value in determining gene function, the need to use later, more genetically stable generations of these mice means that sometimes mice are bred, knowing that 9/10 will be immediately euthanised if they do not have the correct genotype (see Box 6.2). This is a difficult ethical question – the means to determine the result you need can be considered unethical. However, this is precisely the decision made in farming – baby chicks are euthanised immediately if they are male, but kept as egg layers if they are female. Should science and farming be held to the same ethical regulations? As with many ethical questions, there are no easy answers and it is important to make use of institutional animal ethics review boards, animal ethics legislation and animal welfare legislation.

BOX 6.2   CHANGE IN THE NUMBER OF ANIMALS USED IN TESTING

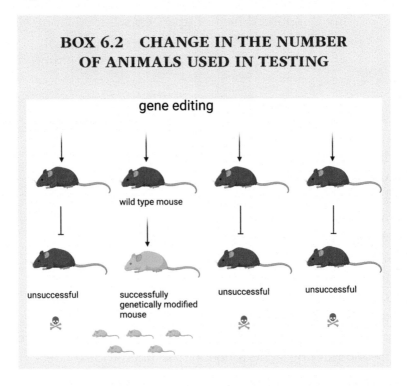

Over the last 50 years, the number of animals used in testing has decreased. This is in part due to stricter regulation by ethical committees and the development of new technologies that reduce the need to use animals. However, since 2012, numbers have been on the rise in some countries again. This is at least in part due to large gene-screening based experiments, where many baby animals are born, tested for genotype and killed if they are not a useful research tool. This includes mutagenesis experiments and CRISPR gene editing experiments *in vivo*. Many animals are not viable post modification, or do not have the required gene modification. This is a new area of animal ethics that will require significantly more consideration as these gene editing tools evolve.

Standardising Animal Experimental Models

The need to perform accurate and reproducible research using animals means that standardisation of models and procedures should be considered. When testing new therapies, then making use of established models is useful because it allows comparisons of multiple studies from multiple researchers. In addition, accurate reporting of the methodologies used is essential for others to replicate the work meaningfully. Standard operating procedures are often used across an institution, but many of these could be adopted worldwide to reduce the variability and thus the research progress of animal experiments.

Human Research

There are numerous ethical issues when performing human research. The most difficult area to address is that of consent. Participants must be fully able to consent to the use of their body or tissues for research. While this seems straightforward, there are several areas of difficulty – How much information do participants get? Is it communicated in the right language and style of language to guarantee informed consent? If giving consent in a hospital setting, there is a risk that patients cannot distinguish between necessary treatments and experimental ones. Cultural safety is essential – many cultures have values embedded in identity and handling of tissue, including data that is collected from trials. The role of ethical oversight boards and committees is essential.

Translation into Human Studies

If a vaccine is based on preliminary rodent and primate studies that show a protective effect, but the vaccine is ineffective in human populations in a clinical trial, researchers may come under fire for rushing into clinical trials when the animal data is not necessarily conclusive. Researchers can face a real dilemma – at what point do you continue with successful animal research or move into clinical trial? The push for faster movement into clinical trials was made apparent in the African Ebola epidemic in 2016–17 – vaccine designers were criticised for not moving into human testing quickly enough. In addition, the ethical questions around the use of more primates need to be considered.

The COVID-19 pandemic led to rapid development of vaccines and therapeutics. Vaccine developers tested the safety and efficacy of candidates in different populations, including large cohorts in many countries, some of which were low- to middle-income countries (LMIC). While the need for an effective vaccine was clear, there are ethical questions raised about the testing: How informed are participants from LMIC? Is there a pressure to participate in a trial when your friends and neighbours are suffering or dying? How many animal experiments were needed to study the disease, or was it feasible to move into human trials quickly? Is it ethical to use live challenge of the SARS-CoV-2 virus to test vaccines?

Testing on Healthy People

Clinical trials of interventions usually move through the phases shown in Figure 6.1. A preclinical phase, which usually, but not always, involves animal testing. Phase 1 trials test the intervention in a small number of otherwise healthy people and look for issues of safety, but not efficacy. Phase 2 trials are usually conducted on the target population, looking for efficacy. Phase 3 trials are usually done on a much larger cohort of people and may include both healthy and diseased/at risk people. Phase 3 trials are usually needed before regulatory approval or authority is given. Phase 4 trials are data collected once the intervention is in widespread use, and so can take years to collect, but often detect issues once many hundreds of thousands of people are involved.

This robust phased method protects individuals without losing the ability to test the intervention. However, the ethics of testing on otherwise healthy people can be difficult. If a disease is widespread, then vaccination of healthy but at-risk populations is easier to justify, for example, Ebola vaccine in DRC, HIV trial in Thailand and COVID-19 vaccine trials worldwide. However, ultimately the subjects undertake some risk with any new intervention. Testing on healthy people also raises issues of *which* healthy people. Most medical testing to date has been performed primarily with male subjects because of perceived and real issues with treating potentially pregnant women. However, this has led to a world of medication optimised for men, and not for women; in many situations, interventions are simply not available for women, limiting their healthcare options, simply because they haven't been tested.

Consent, Coercion and Rewarding Participants

In many countries, participants are paid for clinical trials, which raises issues of abuse of disenfranchised or poor people, and the scientific risk that these people enrol in multiple trials. There are issues around coercion of subjects to be tested. Historical anecdotes of researchers testing their drugs on themselves, or family members, make for great entertainment in lectures or seminars but fail to consider the choices of people involved in early trials. The much celebrated first vaccine, for example, was administered by Edward Jenner on eight-year-old James Phipps, who was certainly not capable of providing informed consent. The discovery of the bacterium, *Helicobacter pylori*, as the causative agent of stomach ulcers was made by Barry Marshall, who tested the theory on himself by swallowing a culture of the bacteria – is this ethical?

In medical research, clinical staff who deal directly with patients are often those who recruit patients to participate in trials or donate tissues. Generally, ethics regulators request that those recruiting have no overall contribution to the individual's care, since again the patient may feel coerced by a doctor they trust or feel that they may be missing out on a potentially beneficial drug that could impact their current treatment success. During recruitment, potential participants are given a patient information sheet that describes the research in lay terms, as well as potential risk and any data or tissue protection. The consent form and patient information form are required to be in appropriate languages and to be culturally informed for a particular population.

Cultural Consideration of Tissues, Data and Future Use

For many cultures, identity is formed through DNA and/or body tissues that have special significance. The idea of cultural ethics has taken a while to become mainstream, but fortunately is now happening. Patient information and consent forms need to consider the diverse population of participants and be able to inform them of all relevant concerns. This includes language, but also explaining how tissues can be stored, used and disposed of in a culturally appropriate way.

Linked to these cultural considerations are concepts of data protection and future use of tissues. Future unspecified use needs to be explicitly stated in consent forms – this is effectively stating that a participant is relinquishing control of their tissue and the associated data that comes from it, forever. It means that researchers can perform other research on the tissue at a future date, even if it is different from the original research proposed on the consent form. Many individuals have no problem with this, but others have beliefs or views that would limit future use. Therefore, an important part of this ethical process is to explain carefully everything that could or could not happen and provide participants with options. Similarly, with the acquired data, there are issues around sharing data, even anonymised, with other researchers. This is commonly done when creating genomic datasets that are shared internationally with other groups, as well as providing access to raw data files when publishing. There needs to be secure storage of both tissues and the data that come from it. All this needs to be considered to provide the best information to participants and allow them to choose with full knowledge.

Conversely, it is possible that new data is obtained from tissues or studies that were outside the original scope of the study design. For example, new DNA sequencing technologies mean that extensive information on molecular tumour types can be found in tissues collected 30 years ago. If this tissue is analysed with new techniques, it is possible that familial cancer susceptibility genes or mutations could be found that would affect descendants. What is the process for dealing with such incidental findings? Ethics regulatory bodies may require a separate process for dealing with incidental findings, but this can never be simple or straightforward.

Regulation of Ethics

Any new experiment or project involving animal or human participants is usually reviewed by a panel of individuals to assess the ethical implications of the research and the plans in place by the researchers to deal with ethical issues. For animal research, this is usually managed at an institutional level, whereas human research can be managed locally, or on a larger, national scale, to ensure robustness and fairness. Ethical panels usually include science specialists, clinicians/vets, non-scientific people and cultural representatives. Ethics applications are important and the value of a review panel in maintaining ethical standards and consistency can't be over emphasised. Every journal that publishes research requires a statement of, and often proof of, ethical approval by the relevant regulatory body.

Responsibilities of the Researcher

The ethics of science is not always straightforward, and researchers bear much of the responsibility of becoming informed and acting in a professional and ethical manner. For new researchers, one of the first ethical issues to grapple with is around honesty and omission of data. How ethical is it to omit an outlying data point? If the experiment changed slightly between repeats but you believe it had no effect, should you still declare the differences? As with most things, honesty is the best policy. If you aren't sure, then be completely honest and allow others to draw conclusions based on *all* available data and they can include their own biases and preferences.

Conflicts of Interest

At various points in science, researchers may be influenced by outside factors. It is well known that published drug company sponsored clinical research is more likely to give a positive result than similar research funded elsewhere. This isn't necessarily devious, but it is important to provide the reader with information of a *potential* conflict of interest and allow others to determine its effect. As scientists, our personal views can also affect how we perceive the validity or importance of science. Conflict of interest statements are required when writing papers, presenting data, applying for grants, as well as reviewing the work of others. Examples of common conflicts of interest include someone who works in the same department, a personal friend, a parent of your child's friend, a previous supervisor or employer, a previous student or someone who you think is an idiot. As above, the declaration is key – others can decide whether it represents a significant conflict or not.

Summary

An understanding of ethical considerations science is essential. However, this is a complicated field and is sometimes difficult for students to navigate. Fortunately, there are lots of active researchers in ethics and bioethics, and lots of expert advice available. There are also several books and articles appropriate for undergraduate students. It is also important to acknowledge that good ethics can be a very personal definition, and this is influenced by culture, upbringing and the research environment in which you find yourself. The important thing is to always ask for advice if you feel uneasy or if you think you may have inadvertently breached an ethical regulation.

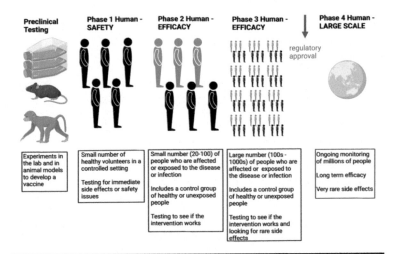

Preclinical Testing	Phase 1 Human - SAFETY	Phase 2 Human - EFFICACY	Phase 3 Human - EFFICACY	Phase 4 Human - LARGE SCALE
Experiments in the lab and in animal models to develop a vaccine	Small number of healthy volunteers in a controlled setting Testing for immediate side effects or safety issues	Small number (20-100) of people who are affected or exposed to the disease or infection Includes a control group of healthy or unexposed people Testing to see if the intervention works	Large number (100s - 1000s) of people who are affected or exposed to the disease or infection Includes a control group of healthy or unexposed people Testing to see if the intervention works and looking for rare side effects	Ongoing monitoring of millions of people Long term efficacy Very rare side effects

Figure 6.1 Clinical Trial Design. These phases are used sequentially to test medical interventions, including vaccines. Preclinical work refers to the fundamental scientific work that creates a potential drug or treatment. Regulatory approval for a new intervention is usually dependent on having studies testing the intervention in animals – often rodents such as mice, but sometimes in monkeys. Once there is a proven effect and safety, then the intervention is tested in a small cohort of healthy humans – this is primarily to test for safety of the intervention and is performed in a highly controlled setting. Phases 2 and 3 are proper randomised clinical trials, where the intervention is tested on increasing numbers of healthy people as well as people for whom the intervention is intended. After these phases, the intervention is granted regulatory approval and sold. Phase 4 refers to the ongoing collection of data around efficacy and safety. This phase can involve millions of people taking the intervention as part of their treatment programme, and adverse effects are collected across all of them – this allows the detection of very rare events that may not have been detected earlier. *Figure created in BioRender.*

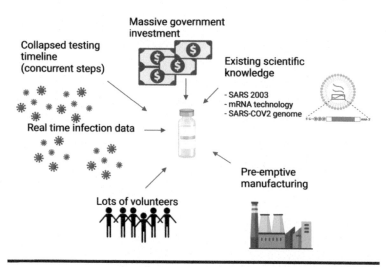

Figure 6.2 Fast Tracking Emergency Vaccines. The COVID-19 pandemic led to rapid development of vaccines. These emergency vaccines can be developed more quickly due to a variety of factors. In the case of COVID-19, there was substantial investment by governments, allowing more research to be done by more researchers. The pandemic was actively infecting millions of people, providing large numbers of willing volunteers who wanted to be part of the clinical trials – usually recruitment of patients is a long and slow process. Similarly, data were collected on the infection in real time, including the sequences of different strains, allowing a huge amount of information to be collected during the trials. There was unprecedented and rapid sharing of scientific knowledge across multiple countries, which also built on existing knowledge from previous SARS outbreaks and existing mRNA vaccine platforms. Finally, governments and companies began manufacturing the vaccines ahead of approval to reduce the lag time for delivery. Other examples of emergency vaccines that were fast tracked primarily due to similar factors include those developed and distributed for the Ebola outbreak in 2013–16, swine flu in 2009 and meningitis

A in 2010. For these emergency vaccines, speed is essential, but safety cannot be and is not compromised. The distribution of successful COVID-19 vaccines is another ethical question. The COVAX programme was established to ensure equitable delivery of vaccines to all parts of the world, but this was not as successful as it could have been, due to nationalistic priorities. *Figure created in BioRender.*

Example 6.1 Failures of Animal Testing

One of the most significant failures of animal testing was the thalidomide incident in the 1960s. People taking the sedative, thalidomide, gave birth to babies with malformed limbs. How did a dangerous drug get released into the human population? Did animal models not replicate this effect? Later studies showed that animals also gave birth to deformed young, but the animal testing experiments were not on the appropriate species or tested on pregnant animals. This controversy highlights the need not just for animal research, but for *effective and relevant* animal research.

Example 6.2 Uninformed Consent

In 1951, a woman named Henrietta Lacks had surgery for cervical cancer. Unknown to her, the medical team collected some of her tumour cells for their own research. They also sent the cells to their research collaborators, who turned these primary cells into a cell line, that could be propagated indefinitely. These HeLa cells were used to make thousands of scientific discoveries and were shared and sold among the research community. However, no one had sought permission from the donor, Henrietta Lacks. As recently as 2021, her descendants filed a lawsuit against a biotechnology company who were still selling the HeLa cells for profit.

In 1964 the Helsinki Declaration was first adopted by the World Medical Association as a set of guiding principles for the ethical treatment of human subjects in research. Since then, the Helsinki Declaration has undergone amendments and updates to ensure that study participants are treated with respect and given all information appropriate for the study population to provide informed consent for the treatment.

Potential Careers

Much of life sciences research involves using animal models of disease. People with an interest in ethical use of animals in research can often become involved in animal research facilities.

Animal Facility Workers

In research animal facilities, there is obviously a role for veterinarians. However, there are also diverse roles including veterinary nurses, animal technicians, technique specialists (for example, surgery), facility managers and animal welfare staff. These roles often require a background in basic science, and sometimes research experience. Those science graduates who care about animals often find these roles rewarding, as they can be an advocate for animal rights at a local (for example, university) or national/international stage. The skillsets required include research design, to help researchers work out how to best complete their research, communication to the public, policymakers and to researchers, and even active research on ethical issues in science.

A good resource for those considering roles in animal welfare is the International Council for Laboratory Animal Science (https://iclas.org).

Resources and Further Reading

Resources Used in This Chapter

- The "Bird in a Vacuum" experiments
  https://en.wikipedia.org/wiki/An_Experiment_on
  _a_Bird_in_the_Air_Pump
- Numbers of animals used in scientific research
  www.theguardian.com/news/datablog/2012/jul/
  10/animal-testing-risk-suffering
- WHO development programme for vaccines
  www.who.int/news-room/feature-stories/detail/
  how-are-vaccines-developed
- Translation into human studies, example of a TB
  vaccine trial
  (Several articles exist and are involved and can
  be searched; this link is the response of the lead
  investigator to the controversy)
  www.bmj.com/content/360/bmj.j5845/rr
- The story of *H. pylori* and Barry Marshall's
  self-testing

Marshall B. *Helicobacter pylori* – a Nobel pursuit? *Can J Gastroenterol.* 2008 Nov;22(11):895–6. doi: 10.1155/2008/459810.

■ The story of thalidomide www.sciencemuseum.org.uk/objects-and-stories/ medicine/thalidomide

Additional Websites That Might Be Useful

■ The development of medical interventions that no longer need animals – an example of pregnancy testing https://sitn.hms.harvard.edu/flash/2018/pee-pregnant-history-science-urine-based-pregna ncy-tests/

■ The COVAX programme for equitable distribution for COVID-19 vaccines www.who.int/initiatives/act-accelerator/covax

■ *Invisible Women* – a book discussing how research and development often only incorporates a male perspective and why this is a problem https://carolinecriadoperez.com/book/invisi ble-women/

■ A podcast about Henrietta Lacks and the HeLa cell line www.npr.org/2013/02/18/171937818/immortal-cells-of-henrietta-lacks-live-on-in-labspo

■ A video about the historical issues in biomedical research www.youtube.com/watch?v=4aFL6HEEouo

▪ Information on Nature Portfolios requirements on ethics for publication www.nature.com/nature-portfolio/editorial-polic ies/ethics-and-biosecurity

Research Publications for Further Reading

1. Expediting the COVID-19 vaccine development

Krammer F. SARS-CoV-2 vaccines in development. *Nature.* 2020;586:516–527. https://doi.org/10.1038/s41 586-020-2798-3

2. Ethics of COVID-19 vaccine distribution

Jecker NS, Wightman AG, Diekema DS. Vaccine ethics: an ethical framework for global distribution of COVID-19 vaccines. *J Med Ethics.* 2021 Feb 16:medethics-2020-107036. doi: 10.1136/ medethics-2020-107036.

3. Emergency vaccines

Ebola – Higgs ES, Dubey SA, Coller BAG, Simon JK, Bollinger L, Sorenson RA, Wilson B, Nason MC, Hensley LE. Accelerating vaccine development during the 2013–2016 West African ebola virus disease outbreak. *Curr Top Microbiol Immunol.* 2017;411:229– 261. doi: 10.1007/82_2017_53.
Swine Flu – Borse RH, Shrestha SS, Fiore AE, Atkins CY, Singleton JA, Furlow C, Meltzer MI. Effects of

vaccine program against pandemic influenza A(H1N1) virus, United States, 2009–2010. *Emerg Infect Dis.* 2013 Mar;19(3):439–48. doi: 10.3201/eid1903.120394.

4. Nuremberg Code – regulating medical ethics and human experiments

Shuster E. Fifty years later: the significance of the Nuremberg Code. *N Engl J Med.* 1997 Nov 13;337(20): 1436–40. doi: 10.1056/NEJM199711133372006.

5. The ethics of biospecimens in research

Beskow LM. Lessons from HeLa Cells: The ethics and policy of biospecimens. *Annu Rev Genomics Hum Genet.* 2016 Aug 31;17:395–417. doi: 10.1146/ annurev-genom-083115-022536.

Use of CRISPR editing in mice
Liu B, Jing Z, Zhang X, Chen Y, Mao S, Kaundal R, Zou Y, Wei G, Zang Y, Wang X, Lin W, Di M, Sun Y, Chen Q, Li Y, Xia J, Sun J, Lin CP, Huang X, Chi T. Large-scale multiplexed mosaic CRISPR perturbation in the whole organism. *Cell.* 2022 Aug 4;185(16):3008– 3024.e16. doi: 10.1016/j.cell.2022.06.039.

6. Working out conflicts of interest

Akl EA, Hakoum M, Khamis A, Khabsa J, Vassar M, Guyatt G. A framework is proposed for defining, categorizing, and assessing conflicts of interest in health research. *J Clin Epidemiol.* 2022 Sep;149:236– 243. doi: 10.1016/j.jclinepi.2022.06.001.

Exercises

Individual

Exercise 1

Rodents and non-human primates are often used for animal research. Cephalopods, including octopuses, are also used. Why is there controversy over the use of octopuses in research?

Exercise 2

Here is an example of 10 statements from a (modified) patient consent form requesting donations of tissue samples from the intestine from people undergoing colonoscopy. The patients have been given a Patient Information Form outlining the details and purpose of the research. The tissue will be used to generate three-dimensional organoids for drug screening.

Consider the purpose of each statement. What other concepts could or should be included?

1. I have read and I understand the information sheet dated 29th October 2022 for people taking part in the study designed to generate lab-based organoids to study the effect of different drugs on the development and symptoms of Celiac Disease. I am satisfied with the answers I have been given.
2. I consent to participate in this study. I understand that taking part in this study is voluntary and that

I may withdraw from the study at any time. This will not affect my continuing health care.

3. I understand that participation in this study is confidential and that no material which could identify me will be used in any reports on this study. Any data collected will be stored securely and anonymised.

4. I understand that consent for colonoscopy is independent of me consenting for this study.

5. I know whom to contact if I have any concerns regarding the study or wish to withdraw from the study.

6. This study has been given ethical approval by the Health and Disability Ethics Committee. This means that the Committee may check at any time that the study is following appropriate ethical procedures.

7. I have consented to my tissue being sent overseas for further analysis.

8. I would like to be informed of the results of the study.

9. I have consented to my tissue being used in experiments with tissues from other people.

10. I have consented to my tissue being destroyed at the end of the study

Class

Exercise 3

Discuss the ethics of using parabiotic mice in experimental research.

Kamran P, Sereti KI, Zhao P, Ali SR, Weissman IL, Ardehali R. Parabiosis in mice: a detailed protocol. *J Vis Exp*. 2013 Oct 6;(80):50556. doi: 10.3791/50556.

Exercise 4

Find a historical or contemporary example of ethically dubious reporting or science based on race or cultural differences. How has the research changed over the last 100 years? What could be done to improve the processes around such reporting and/or research?

Chapter 7

Science Funding

Introduction and Scope

Scientists require funding to complete experiments and share results with the greater community. This funding can be procured from many different sources, with diverse requirements for each application. This chapter will describe how scientific funds are sourced for different purposes, and the application process in general terms.

As students, you are unlikely to be applying for typical government or industry funding yet, however, all scientists involved in experimentation and reporting should have an appreciation for the variety of funding sources and the complexity of application processes and reporting.

DOI: 10.1201/9781003326366-7

Learning Objectives

- To gain an appreciation for the different types of funding
- To describe the general components of a grant application
- To produce a grant proposal based on the criteria and requirements of the national funding agencies in the country where you are conducting your research

Funding for Research

Scientists must be able to pay for their experiments; this not only includes consumables and reagents to perform experiments, but also salaries for staff, infrastructure costs, costs for animals, and regulatory and health and safety requirements, animal ethics, patient recruitment and so on. Increasingly, costs for data analysis, management and storage must also be considered.

Because scientists work in a variety of settings, with diverse employment situations, acquisition of funding can become very complicated. Generally, there are four sources of funding:

1. Government (for example, National Institutes of Health (NIH; USA))
2. Industry (for example, pharmaceutical companies)

3. Charity (for example, Bill and Melinda Gates
   Foundation (BMGF))
4. Employer (for example, university)

Many of these sources require a competitive bidding
process because resources are usually less than what
is required. Often, multiple stages of bidding are
required.

Sources of Funding

Granting agencies can provide money from amounts
as low as hundreds, to as high as millions or even
billions of dollars. The decision to apply for a
particular pool of money depends on the scope
of the project, the amount of evidence required
to show feasibility, the type of research and the
timeframe proposed in which to complete the
work. Grant writing requires considerable time. An
editorial in *Scientific American* found that faculty at
academic institutions spent 40% of their time on the
bureaucratic and administrative side of research –
finding and applying for different funding sources
(www.scientificamerican.com/article/dr-no-money/).
This 40% didn't include the time it takes to actually do
the research!

Government

Government funding is, in theory, supposed to
provide large amounts of money to fund research that

is seen to contribute to the public good. This includes both research with a defined application, such as a new cancer drug, but also research that provides fundamental knowledge about the world, without the requirement to show an immediate output (such as the structure of a protein or chemical). All science research can be considered valuable, but many government funding agencies have a specific purpose. In biomedical research, this usually covers basic, clinical and applied research with a health focus and can often be divided into subject areas, for example the National Cancer Institute or the National Institute of Diabetes and Digestive and Kidney Diseases within the National Institutes of Health in the United States.

It's important to note that most government funding is supported through taxpayer dollars and indicates that the research is publicly funded. Many public funding bodies require evidence of response to a country's particular needs or a discussion of research impact – how the research may benefit society, the economy, the workforce, patients, etc. Most of these funders also require the research to be published under open access, or that the data are available in public repositories so that the results of the investments made by taxpayers can be viewed by those same taxpayers.

Industry or Contracts

Industry (or contract) funding sometimes has a negative connotation because of the perceived lack of freedom in designing and performing research and/or restriction of data generated from the research. Most companies have their own research and development teams but recognise the need for collaborative and interdisciplinary research and so form relationships with researchers external to the company. In fact, some companies have government funding or contracts with public funders and require collaborative agreements to move projects forward with expertise from academia or non-profit research institutes. Other collaborations between industry and academia can be funded through small business research grants or venture capital sources.

Non-Profit Organisations

Charitable funding for biomedical research is usually sourced by registered charities that collect funds from individuals or organisations to specifically target research towards a disease or treatment. Each charity funding service is different, but often there is an expectation of research that has a clinical outcome at some point. However, many charities also fund fundamental research. Funding for individuals, via fellowships, is also common, and this support allows researchers to develop or expand their own research programme. Reporting of outcomes is particularly important for research funded by charities – the line from donor to outcome is more direct than via government funding via tax collection, and donors like to see a focused and meaningful outcome.

Employers

Employer funding usually refers to universities or institutions that provide internal funding, often for new researchers. These funds are usually designed to kickstart research programmes, or to invest in exciting research results that may have a positive impact on the institution and its reputation. These internal grants can be a great source of revenue to test new strategies and to collect pilot data before applying to an external funder for money. Institutes often fund parts of the pathway to commercialisation of a research finding.

BOX 7.1   RESEARCH IMPACT

Research impact is sometimes requested by funding bodies – this refers to the line of sight of the research proposed to an impact on society, health, the economy and/or the community as a whole. For example, will your research lead to a new drug? Will your research lead to a governmental policy change? The *pathway* to impact is how a researcher can show the steps required to make this impact – some of which may be the responsibility of the researcher or the project, or of future researchers. Given that the vast majority of research is funded by the taxpayers, proving the research may have value is important. Most organisations have professional support to help with impact.

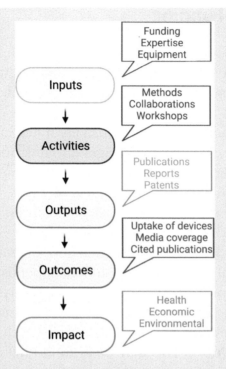

Research impact refers to the pathway from idea to the relevance of the idea: what you need to do the work (inputs), how you plan to do it (activities), what will come from the research (outputs), how those outputs will lead to changes in practice (outcomes) and finally what that will mean in a broader context, e.g., government policies, or changes in medical interventions (impact). *Figure created in BioRender.*

Types of Funding

There are sources of funding for many types of costs related to research. These include the following:

Projects – This refers to discrete research projects designed to answer a question or hypothesis.

Programmes – These are usually much larger pools of money and are designed to link and support multiple related projects together, for example, multiple ways to study the effects of a new drug – one project may involve toxicity studies, another efficacy studies.

Salary – Funding for research support staff, postdoctoral researchers or student stipends, to work on a project. These may be judged based on the scientific project, the quality and potential of the candidate and usually are a mixture of both.

Scholarships – This may be to fund PhD or MSc student living costs or experimental costs, or they may support professional development of staff and students. These are almost exclusively decided based on the quality of the candidate and their intentions for use of the scholarship.

Travel – Effective scientific research involves discussion of ideas and results with other scientists. Travel to international conferences and visits to other laboratories to discuss research, therefore, is essential. Many funding agencies provide small grants to facilitate travel to attend conferences or to support visits to learn new techniques or to present data.

Conferences – Funders can support conference hosting costs, recognising the need to bring researchers together. Many conferences can also be supported by industry donations (for example, suppliers of lab reagents often sponsor discipline-specific meetings) and charitable donations, especially smaller conferences.

Equipment Funding – Funding for specific pieces of equipment, designed to support one project, or an entire institution, for example, an electron microscopy unit.

Seed Funding / Feasibility Studies – Funding for large projects often requires evidence that the project is likely to work. Many funders provide strategic or feasibility funds to allow people to investigate a preliminary idea or concept. This may be to test the timeline of delivering a new vaccine within a health system, to test a new technique or to determine whether a collaboration across continents will actually work.

Clinical Trials – Clinical trials are long, collaborative and expensive. Many funders have designated funds to support clinical trial research with more requirements for proof of need or efficacy.

Funding Processes

Different grant bodies have different processes by which they assess and award funding. It is not possible to cover all the different types of grant applications and processes in this book, rather we will provide an overview of the key components of most funding applications.

The bulk of grant application processes begin with the submission of a research project. The extent and detail vary enormously; however, a clear research question and experimental plan is almost always required. Many organisations have a multistep approach, whereby a preliminary proposal (one to three pages) is required at first submission and then successful applicants from the first assessment are invited to submit a larger, more complete, full proposal. This is an efficient way to judge many projects but has some limitations – the ability to sell an idea in a small amount of space becomes an essential skill for a researcher.

Following receipt of an application, grant projects are usually first assessed by a panel of scientists. Projects that are deemed poorly prepared, scientifically poor or outside the scope of the granting body are rejected immediately. The second phase usually includes scientific review by a panel of expert scientists, who are usually not part of the original assessment panel and are therefore independent of the granting body. These expert reviewers can be recruited from anywhere in the world and the process is designed to get a fair and impartial assessment of the quality and feasibility of the scientific project proposed, judged by someone with the specialised knowledge and expertise to make such a decision. Almost always, these external reviewers perform this task on a voluntary basis but may be given a small honorarium for their time and expertise.

The reviewer assessments are returned to the original panel, and there may or may not be an opportunity for the researcher to respond to the reviewers' comments. The scientific panel makes the ultimate decision on what projects to fund, based on review comments, their own discussion of the merits of the proposal and by the amount of funding they have to distribute. Because of limited resources, many scientifically valid and interesting research projects may remain unfunded, or partially funded, simply because there is not enough money to fund all deserving projects.

Components of a Funding Application

The overall components of a grant application may consist of some, or all, of the following:

1. Project design, including preliminary or pilot data
2. A description and justification of the people involved in the research, including collaborators with resume/CV from all primary investigators
3. The facilities / resources to be used, and support from institutions to guarantee access to technology and expertise
4. Budget
5. Impact or significance – proof that this research is needed; similarly, proof that the project aligns with the strategic goals of the funder

6. Ethical and regulatory approvals – proof that appropriate ethical approval has been given; this also includes any cultural considerations that may be required; and any health and safety or other legal requirements (for example, regulations concerning genetic modification)
7. Proof of access to required facilities, equipment, etc.
8. Letters of support from collaborators, guaranteeing scientific and potentially budget support

BOX 7.2   RESEARCH OVERHEADS

Most research is performed in institutions. These institutions have running costs, like buildings, electricity and administrative staff. They can recoup some of these costs by charging overheads on research funding. This means that a proportion of the grant awarded will be paid to the institution to support these costs and allow the researcher to do the experiments in the grant proposal in a real lab with power and gas for their Bunsen burners. However, there are no firm rules about overhead rates, with some institutions charging over 100%, or applying overheads to some, but not other, charges or individuals. While overheads are essential, they can often have a massive impact on the grant budget, thus limiting what experiments can be done, or whether the grant is viable at all. Many early career researchers, including their own salary costs in the project, can find these costs prohibitive.

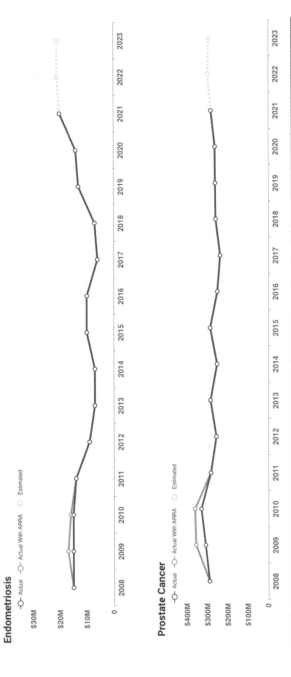

Figure 7.1 Funding Sources and Expenditure. The National Institutes of Health fund health related research in the United States. The amount that is spent on each disease varies year by year but is a useful tool to see what government health priorities are at any given time. The figure below shows spending on

endometriosis (top, affecting 6.1% of U.S. women) versus prostate cancer (bottom, affecting about 12% of U.S. men). Data like these are useful tools for working out where patient advocacy groups can concentrate their efforts. https://report.nih.gov/funding/categorical-spending#/

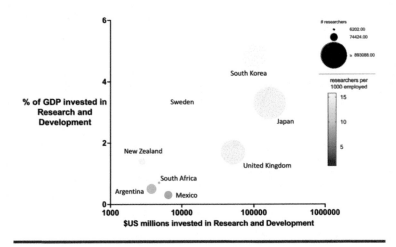

Figure 7.2 Funding of Research and Researchers. The figure shows how much of the gross domestic product (GDP) each of selected countries invest in research and development. Note that this spending is all research, not just life sciences. It is interesting to compare across nations and discuss how and why these decisions are made. The figure also shows the number of people in the research and development workforce – is there a link between investment and careers? Why do you think some of these don't seem to match? Many researchers spend part of their careers involved in helping governments decide research priorities or lobbying for increased investment. It is important to understand how these decisions are made and to work out what influence a research community may have in affecting these decisions. Data from OECD, 2018–2020, https://data.oecd.org/rd/researchers.htm#indicator-chart

Example 7.1   Early Career Funding

Many granting agencies/foundations provide specialised funding for early career researchers (ECR). It is important to check eligibility for each ECR award as well as to determine the best fellowships for an individual circumstance. Here we compare two award programmes from our two different countries. Note the differences in eligibility, application details, budget and timeframes.

	New Zealand – Marsden Fast Start	United States – NIH Ruth L. Kirschstein Predoctoral Individual National Research Service Award
Goal of programme	- The Fast-Start programme is targeted at researchers who are employed at New Zealand universities, Crown Research Institutes (CRIs) and other research organisations, and who are trying to establish independent research careers and create research momentum	- The purpose of this Kirschstein-NRSA program is to enable promising predoctoral students with potential to developinto productive, independent research scientists, to obtain mentored research training while conducting dissertation research. - This programme is intended for people obtaining their PhD and is designed as a training programme on the path to independence.

	New Zealand – Marsden Fast Start	United States – NIH Ruth L. Kirschstein Predoctoral Individual National Research Service Award
	- This programme is intended for people early in their research careers. - *The applicant should be involved in their own independent research, and not merely be part of a larger group's research programme.*	
Eligibility	- Up to 7 - Any - Must have an employment contract with an institution	- Second or third year predoctoral student - U.S. citizen or permanent resident, enrolled in a research doctoral degree program. - Must be enrolled in a degree program at U.S. or foreign institution.

	New Zealand – Marsden Fast Start	*United States – NIH Ruth L. Kirschstein Predoctoral Individual National Research Service Award*
Application process	Expression of Interest (one page) Invited to Full Project application (10 pages)	Application includes both research and training plan and is designed to be written with input from mentor. Page limits on individual sections
Conditions of grant - Length - Amount - Country of research	3 years $120,000/year Must be based in NZ	Up to 3 years Funds to cover tuition and fees U.S. or foreign institutions
Assessment criteria	- Potential for significant scholarly impact because of the proposal's novelty, originality, insight and ambition. - Rigorous and should have a basis in prior research and use a sound research method.	Overall impact assesses the likelihood that the proposed training will enhance the candidate's potential for a productive, independent research career.

	New Zealand – Marsden Fast Start	United States – NIH Ruth L. Kirschstein Predoctoral Individual National Research Service Award
	- Research team must have the ability and capacity to deliver. - Should develop research skills in New Zealand, particularly those at the postdoctoral level and emerging researchers.	
	- Must consider the relation of the research to the themes of Vision Mātauranga and, *where relevant*, how the project will engage with Māori.	– Applicant: commitment to independent career, evidence of productivity – Sponsor/ collaborators: qualifications, mentoring record and defined role for collaborators

	New Zealand – Marsden Fast Start	United States – NIH Ruth L. Kirschstein Predoctoral Individual National Research Service Award
		– Research training plan: quality, significance, impact and rigor of proposed research during the training period. – Training potential: individualised and mentored experience to develop knowledge, research and professional skills.
		– Environment and commitment to training: are facilities, resources and opportunities (seminars, workshops, professional development) adequate and appropriate?
Success rate	13% in 2021	26% in 2021

Example 7.2 Intellectual Property and Patents

Any research has the potential to lead to a commercial application. This means that researchers need to consider the associated intellectual property (IP) and patent regulations. Most institutions, such as universities, have IP policies, as well as specialised staff to work with researchers. However, many researchers do not understand the pathway from basic idea to commercialisable product. Researchers have a responsibility to funders as well as their institution to be aware of when they can talk about their research, and with how much detail. Intellectual property (IP) can be particularly difficult for students to navigate – a staff member at a university for example, is being paid to work there. A student is paying to do their work. How to reconcile the obligations to the host university with the intellectual outputs? How does the funder's contribution affect IP? If a company has invested, don't they deserve some financial reward? But what if a charitable organisation or a government fund the research? What are their policies for IP? IP and patent legislation and policies can vary widely between institutions and in different countries.

Potential Careers

There are many ways to acquire funding for research. Conventional grant applications are the most well-known, but an important source of funding is industry and venture capital businesses, companies who want to provide funds to research because it is a financially sound investment.

Venture Capital Investments

Many venture capital companies, who provide funds to support research projects and development of new technologies and products, rely heavily on scientific expertise to assess applicants and to identify important new areas of research. Science graduates with a diverse range of specialties and interests are in high demand in these companies. These roles are well suited for those who like to see real-life and real-time outcomes from research, as well as for those who enjoy the scientific aspects of economics. Key skills are critical thinking, communication and attention to detail.

Resources and Further Reading

Resources Used in This Chapter

■ Biotech and investment in research

Silverman E. Follow the money: How biotech's funding shapes its direction. *Biotechnol Healthc*. 2004 Jul;1(3):41–5.

■ Marsden Fund Fast Start (NZ)
www.royalsociety.org.nz/what-we-do/funds-and-opportunities/marsden/marsden-fund-application-process/submitting-a-proposal/preliminary-propo
sal-guidelines-for-applicants/

- NIH Predoctoral fellowship (USA)
  https://researchtraining.nih.gov/programs/fell
  owships/F31
- How much time is spent sourcing funds
  www.scientificamerican.com/article/dr-no-money/)
- Category-based spending from NIH
  https://report.nih.gov/funding/categorical-
  spending#/

Additional Websites That Might Be Useful

- Bill and Melinda Gates Foundation
  www.gatesfoundation.org
- A discussion of philanthropy as a source of
  research funding
  www.nature.com/articles/d41586-023-00077-2
- How to understand and apply research impact –
  a guide from the University of Auckland, New
  Zealand
  https://research-hub.auckland.ac.nz/subhub/resea
  rch-impact
- Understanding overheads costs in research
  www.srainternational.org/blogs/srai-jra1/2021/11/
  22/overhead-rates-impact-on-research-applicati
  ons-suc'
- Springfree trampolines as a case study in research
  and intellectual property
  www.stuff.co.nz/business/unlimited/innovation/
  9989660/Trampoline-firm-gets-jump-on-competit
  ion; www.springfreetrampoline.com/blog/2019/
  04/01/how-i-built-guy-raz-springfree-trampoline/

▪ Example of a university Intellectual Property policy for students www.otago.ac.nz/administration/policies/otag o003228.html

Research Publications for Further Reading

1. Research Impact

Penfield T, Baker MJ, Scoble S, Wykes MC; Assessment, evaluations, and definitions of research impact: A review. *Res Eval.* 2014 January;23(1):21–32, https://doi.org/10.1093/reseval/rvt021

2. Funding and research outputs

Webster EM, Jensen, PH, Clark J, Hirsch G; Research funding mechanisms and biomedical research outputs (October 29, 2021). 7. Clark J, Hirsch G, Jensen P, Webster E. Research funding mechanisms and biomedical research outputs. *Econ Pap.* 2016;35:142–154, Available at SSRN: https://ssrn.com/abstract=3952 254 or http://dx.doi.org/10.2139/ssrn.3952254

3. A journey of funding – HIV vaccines

Shapiro SZ. HIV Vaccine development: 35 years of experimenting in the funding of biomedical research. *Viruses.* 2020;12(12):1469. https://doi.org/10.3390/ v12121469

Exercises

Individual

Exercise 1

Choose one of the following research areas and create a research impact pathway:

- Chemical structure of a new chemotherapy drug
- Analysis of soil microbiome in intensive farming
- Review of housing conditions and health of children

Exercise 2

Create a list of criteria you would use to assess the quality of a grant designed to fund a new project by an early career researcher.

Class

Exercise 3

In groups, create a research project, covering background and rationale, methodology and experimental design, and significance. Include justification of design, for example, statistical power. The entire project needs to fit into one page!

Once you have finished, switch proposals with another group and provide feedback as grant reviewers.

Exercise 4

Compare and contrast articles on a topic (such as tobacco research or vaccines) that was funded by a company in the 1950s or 1960s compared to present day. How does reporting by the company and transparency of the work differ between the present day?

Chapter 8

Scientific Publishing

Introduction and Scope

The goal of scientific research is to discover new things – these may lead to changes in how we run our lives, improvements in health or impacts on the environment. New data also provide building blocks for other scientists to further discovery. Rarely does one scientist or project in isolation lead to lasting changes in dogma; it is a step-by-step process with contributions from multiple researchers over time that build together to take the field forward. Every researcher reviews what has been done before and decides where to go in the future. This means science constantly changes, old ideas are retained or rejected, despite good science pointing to a conclusion that was once valid with the knowledge at that time. It is often difficult for non-scientists to view the rejection of once established ideas as proof of scientific

DOI: 10.1201/9781003326366-8

progress. To facilitate these discussions and to direct research, all scientific results need to be shared with the community. While social media and other internet-based dissemination are more and more common, the gold standard for disseminating scientific work is through publication in peer-reviewed scientific journals.

This chapter explains how the publication process works for scientific journals. It also discusses other means of data presentation and some of the limitations of the traditional publishing models. This chapter touches on the rise of misinformation in science, especially in social media and the importance of science communication to the public.

Learning Objectives

- To understand the pathways to scientific publication
- To discuss challenges in publishing data and the current publishing models
- To appreciate the role of social media in disseminating information and the responsibility of scientists to communicate with the public

Why Publish?

Science only moves forward if other scientists know what has been done. The traditional method for sharing data and results is through publishing research outcomes in scientific journals. A key component of this process is peer review – other scientists in your field read the paper and judge it on quality, novelty and rigour, and then make recommendations to the journal. Once a paper is accepted for publication, it represents a significant achievement for the scientist – their work has been recognised as high quality and will now be shared with the science community. The idea of sharing data publicly means that other researchers in the field can repeat your work as validation and can also build on it to advance knowledge.

For an individual researcher, their research is disseminated in the scientific world through a peer-reviewed publication in a scientific journal. Publications increase the profile of individuals, and the amount and quality of their research publications often have a significant impact on their career and their chances of acquiring research funding for the future.

Scientific Journals and How They Work

Scientific journals are businesses – they are not run by scientists as a public service, rather they are owned by publishing companies and therefore incur a cost in publishing and dissemination. Many professional societies (groups of scientists in a related field, such as the American Society for Microbiology) also run journals, but usually in partnership with a publishing company. In the current model, readers must purchase scientific articles (there are exceptions, see below). An interesting quirk of the scientific publication process is that the scientific decisions and associated work are performed by other scientists on a voluntary basis. One benefit to this approach is that scientists are freed from commercial pressures in determining what is good or not. One drawback is that the most qualified scientist to review another's work may not have time or resources to commit to this important review process.

Generally, journals are specialised in areas of science, focusing for example on cancer genetics, or epithelial physiology, although there are several broader scope scientific journals such as *Nature* and *Science*, or medical-focused journals such as *The Lancet*. These broad scope journals publish across a wide variety of specialties including physics, geoscience, astronomy and computer science as well as cardiology and neuroscience. Even within journals such as *Nature* and *Science*, the last 20 years has seen an explosion in various specialised offshoot journals such as *Nature Metabolism, Nature Genetics* and *Science Translational Medicine*, partly due to the volume of high-impact research sent to these journals for publication.

Different journals vary in the type and format of research they publish. As a general rule, a journal article comprises an Introduction, which outlines the current knowledge and identifies a new research question (see Chapter 2, Box 2.1). A Methodology section then explains the experiments in sufficient detail that they could be repeated by other researchers to validate the findings. The Results section then shows the data in a series of diagrams and graphs as Figures, with supporting text. The last section is the Discussion, which is an opportunity for the authors to state the novelty of the findings, the answer to the research question and to put their results in the context of other people's research and knowledge in the field. Finally, a short summary statement of the entire paper, an Abstract, is attached to the start of the article, and this is often what other researchers look at to determine whether they should read the whole paper. Although the Abstract is positioned at the start of the journal article, it is often written last. Sometimes additional detailed data or methodology is provided as Supplementary Material – it is not part of the research paper but may provide other useful information such as primer design or pilot studies. A research article generally contains between three and ten figures and is usually five to six pages long.

Who Decides What Gets Published?

There must be scientific input into content. Science journals generally have an editorial board, with an editor in chief and section editors responsible for making decisions on which papers are ultimately published in the journal (Figure 8.1). Invitations to sit on editorial boards are another metric of success for scientists and a recognition of their expertise in the field. The editorial boards partly determine the scope

of the journal and are also responsible for selecting
reviewers for the articles, collating the review reports
and, finally, making a decision on whether the article
should be published.

Peer review is the key factor in scientific publishing
and ensures that high-quality science is published.
In this process, an article submitted to a journal
is reviewed by the editors for initial quality and
then sent to at least two independent scientists
with expertise in the field. They read the article
and critique it on scientific quality, appropriateness
of conclusions, clarity of data presentation and,
in many cases, novelty and impact of the work.
Peer reviewers generally write a detailed report of
positive and negative issues in the article and make
a recommendation about publishing the article. This
is sent back to the editorial board to either reject the
article, or to send the report to the authors to give
them an opportunity to revise the article and bring it
up to the quality needed.

Peer review is a voluntary effort, although there
are new means to publicly record how many articles
someone has reviewed to demonstrate their own
workload in this area (for example, Publons). Peer
review is usually blinded at least one way – the
majority of journals do not reveal the names of
reviewers to the authors, but the author names are
revealed to the reviewer. This creates some bias, of
course, and some journals use double-blind review,
or publish the names of the reviewers alongside the
article once published (for example, Frontiers group),

which means that known biases or conflicts need to be owned by the reviewers.

There are, of course, problems with peer review, and sometimes poor-quality science is published (for example, an article in *Lancet* in 1996, which claimed the MMR vaccine caused autism, later retracted for poor data quality (see Chapter 1)). In some highly specialised fields, it can be difficult to find peer reviewers with specialist expertise who do not have a conflict of interest with the authors or editors. Some peer reviewers are committed to one view of an area and may not be able to provide unbiased review, although this is one reason why multiple reviewers are used. Sometimes it can just be difficult to find enough volunteers with time to provide review. And sometimes peer reviewers do a cursory job, and their review has no value.

Basic standards for peer review can also be difficult to maintain. A scientist may have excellent knowledge in a scientific area, but perhaps not specific knowledge of the experimental method or statistical analyses – in these situations, a reviewer should acknowledge their own lack of expertise (assuming they know they have a lack of expertise) to allow the editors to take that into account.

Some newer journals have different peer review criteria. For example, *Scientific Reports* asks reviewers to critique the experimental work, but not to judge novelty or impact, preferring to publish good quality science and to allow readers to determine impact on their own. This is an example of the changing world of scientific publishing to recognise a diversity of excellence and to remove some of the entrenched biases of the traditional system, which tends to favour established expertise.

Finally, publishers and the scientific community have recognised the value in publishing data that represents a small step in progress or that highlights concepts or techniques that do *not* work. These data have traditionally been excluded from publication, with the belief that only positive results (data that proves a hypothesis) were useful to share. However, this model did not provide scope for all of those failed experiments – those that show a technique is not appropriate for a particular research question, or those that show an intervention has no effect. Publishing such negative results prevents other researchers wasting time and resources repeating experiments that will not work. Some publishing companies specifically created journals to publish such small advances or negative results (for example, *BMC Research Notes*).

Revisions and Refinements

Peer review reports provide a basis for editorial decisions. Editors may reject the article based on quality or scope and suggest publishing in a different journal. They may accept the article for publication. Usually, however, articles require revision and refinement, and authors use the peer review reports to improve the article. This may require extra experiments or including more discussion of aspects of the work. Upon resubmission, the article may be returned to peer reviewers (often the original ones) or reviewed by the section editor only. The editors should be seen as individuals working with the authors to help them create a high-quality article, rather than gatekeepers blocking publication. There is often significant dialogue between authors and editors and between reviewers and editors, to together create a high-quality article to be published in the journal. Remember that the journals are a business, and it is in their best interest to publish work that other people want to read and therefore buy.

Corrigenda and Retractions

Occasionally an article is published that contains errors. These may be as small as mislabelling a figure or as large as misinterpreting a result, thus invalidating major conclusions. It is the authors' responsibility to alert the journal if errors are detected, and the journal can publish a correction. This correction is linked to the article so that searches of the original article also detect the correction – in this way, the correct information is available. Corrections can be seen as a constant refinement of the article. For major issues, such as incorrect conclusions or results presented, a journal may retract an article. In this situation, the article is completely removed from the journal and all online repositories, and it is as if it never existed. This is a serious decision and usually reserved for either problems with technology/methodology or in some cases fraudulent activity.

BOX 8.1   HOW TO WRITE GOOD

Scientific publishing requires researchers to write about their work. This means that an essential skill for a researcher is the ability to write well – clear, simple messages are important, but sloppy writing also implies a lack of attention to detail that could also affect interpretation of the data. If someone can't check the words are spelt right, how careful are they in determining if they have the right experimental controls? Many students have had no formal training in writing well, but some important, easy to check, concepts are listed here:

Paragraphs: A paragraph covers one key point. Each paragraph needs to start with an introductory sentence, then one or more sentences covering one key message and then a concluding sentence summarising the content and potentially leading to the next point / paragraph.

Sentences: Sentences need to be short, simple and direct. Scientific writing is about communicating a clear message, not crafting an elaborate narrative. Do not be too concerned about staccato sentences (lots of short sentences) – these are actually more useful than a long winding sentence joining lots of concepts together.

Words: Always keep language simple. Use the simplest word that is required (for example, "use" not "utilise", "new" not "novel").

Thesaurus-heavy content is more difficult to read than simple text. Remember to check the language requirements of journals – there are a variety of English styles that can be used – the authors of this book spent hours reconciling our contributions written in US-English versus AUS-English.

Punctuation: Incorrect punctuation, such as a comma or apostrophe in the wrong place, can have a significant effect on the meaning of a sentence. Anything that creates confusion is a bad idea in scientific writing, so correct punctuation is important.

Proofread, Don't Spell-Check: While writing software has improved a lot and provides helpful advice, computers will not replace humans for checking context and clarity. Many words are spelt right but are in the wrong context (for example, "lead" versus "led"). Proofreading your own writing and asking others to do so, will vastly improve the quality.

Measures of Journal Quality

In recent years, the number of different journals has expanded hugely, so how to decide where to publish? The traditional model of scientific publishing uses a series of metrics that attempt to determine how important a journal is in the field. This is measured by, for example, the impact factor, which is a number determined by counting how many times articles published in the journal are cited by other articles. This provides a measure of impact of articles in determining which (or whose) science is considered important. There are obvious problems with these metrics, including self-citation (usually excluded from these metrics) and lazy citation, where researchers simply cite articles from a researcher who is renowned in the field, rather than performing a comprehensive and up-to-date review of the existing literature. More recent methods to determine impact include number of downloads/views, or impact via social media. As the scientific publishing industry catches up with modern technologies, it is likely that the reliance on impact factor and associated metrics to determine scientific quality will be reduced, and potentially replaced with real-life impact.

BOX 8.2   ALTMETRICS

Traditional measures of research quality have depended on citations of published work. An article that is often cited by others can be considered to be more important or have a higher impact than one that is cited less often. As the number of research publications has grown exponentially over the last ten years, these metrics are becoming less reliable. It is simply no longer possible to read every paper that may be related to your work, meaning some important papers are missed out, or that often-cited articles are easier to find and therefore are cited more often.

The impact factor of journals is also determined by the number of citations of their article by other articles. These are calculated independently but are used by journals to recruit more high-quality research articles (remembering that journals are a business). Researchers often feel pressure to publish in high impact factor journals, but newer measures of impact are becoming more influential. For example, a researcher could choose to publish their work in a journal with a high impact factor, or a journal that is read most widely, depending on how they want their research to have impact.

Altmetrics refer to the other measures of research impact – they include qualitative data that complement the citation-based metric. Some altmetrics include citations on Wikipedia, incorporation into policy reports, public media coverage and social media presence. There are now established ways to calculate the Altmetric value of a research work, and this can be added to the traditional metrics to assess the impact of the research work. Altmetrics can be very diverse.

The website www.altmetrics.com highlights the value of the following:

> A record of attention: *This class of metrics can indicate how many people have been exposed to and engaged with a scholarly output. Examples of this include mentions in the news, blogs, and on Twitter; article pageviews and downloads; GitHub repository watchers.*

> A measure of dissemination: *These metrics (and the underlying mentions) can help you understand where and why a piece of research is being discussed and shared, both among other scholars and in the public sphere. Examples of this would include coverage in the news; social sharing and blog features.*
>
> An indicator of influence and impact: *Some of the data gathered via altmetrics can signal that research is changing a field of study, the public's health, or having any other number of tangible effects upon larger society. Examples of this include references in public policy documents; or commentary from experts and practitioners.*

New Challenges for Scientific Publishing

Costs of Publishing

Because journals are businesses, they need to make money. They generally do this in two ways – (i) by charging people to read the articles (by subscription or by paying per article) and (ii) by charging authors to cover the publishing costs. Traditionally, articles were published as hard copies in actual books, and publishing costs covered coloured inks and page charges. In a world where most of the access to articles is done online, these charges are becoming harder to justify. Obviously, some costs of administration and promotion need to be covered, but there are very strong arguments that individuals who pay for research via taxes should not then have to pay to read the results. This is an area of rapid change.

Open Access Models

In response to these traditional funding models, many journals offer Open Access articles, either per article or by making their entire journal free for anyone. This resolves the problem of making scientific findings available for all to read, yet the cost for this is usually borne by the authors and can be prohibitively expensive for small researchers without much funding. This in turn creates a publishing bias. As a compromise, many journals release their articles as open access after a period of time, for example, six months, which allows them to make money, but the research is still available in a reasonable amount of time. Fortunately, many funding bodies now include the requirement or at least opportunity for researchers to request funds to publish the findings in Open Access format.

The COVID-19 pandemic is an example of how the traditional publishing model can be adjusted in real time. Because it was crucial for scientists to share their data and for newspapers to report the facts about the virus and disease, many publication paywalls and costs were lifted for journals to provide all COVID-19 manuscripts freely available. In addition, the number of COVID-19-related manuscripts on preprint servers exploded during the pandemic to help keep scientists apprised of real-time changes in disease progression and vaccine trials.

Availability of Data

Research now can include enormous amounts of data and/or significant steps in data processing. While some of these data and processes are available to peer reviewers to assess, they are not always available for readers of a published article to look at and determine whether the data was correctly acquired and analysed. Many journals therefore require the deposit of raw data on online servers and these data are linked to the article. Other journals require a data availability statement, where authors can provide data on request or explain why raw data is unavailable (for example, due to commercial or ethical sensitivity). A common example of this is high throughput sequencing data and omics data that represent a databank of information that can be incorporated into larger datasets to validate results.

Pre-Publications (For Example, BioRxiv)

The peer review process can be very long and time-consuming. This means that important data is not always available to researchers in a timely fashion. While the process generally ensures quality, it can be problematic in a quickly moving field. This was particularly apparent in the first half of 2020 when several studies on COVID-19 and SARS-CoV2 biology needed to be available to all researchers as quickly as possible. Preprint servers are repositories of research articles that have not yet completed peer review. The advantage of these repositories, where authors can pre-publish their manuscript and make it available to anybody to read, is that anybody can read it and use the information. The downside is that the lack of peer review means that high quality and low-quality science is available and distinctions between the two may not be possible. This is potentially dangerous. Finally, some journals will not accept articles that have been pre-published because of the potential of losing the unique opportunity to publish the work. In the case of COVID-19 and SARS CoV-2 research results, most of the BioRxiv or MedRxiv preprints eventually were published in high-quality, peer reviewed journals.

Predatory Publishers

The world of open access publishing and the ease of online publishing has unfortunately led to a proliferation of corrupt publishing practices. This is known as predatory publishing, where journals have been created and publish research without rigorous peer review but at a cost to the author. They therefore earn money for publishing articles but have allowed poor quality science to be published. Many of these journals or families of journals provide false information about peer review processes or editorial boards to give an impression of rigour. These journals

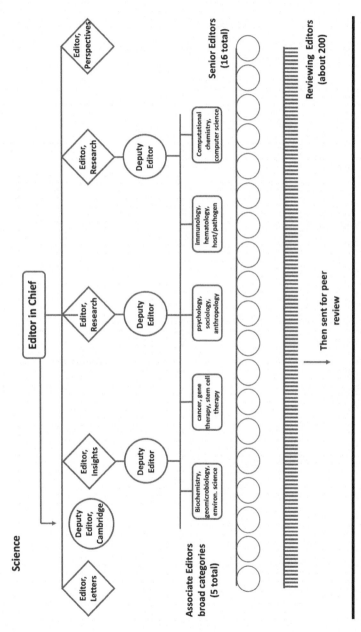

Figure 8.1 Editorial Structure of a Science Journal. An editorial board comprises experts in the specialist area or areas covered by the journal's scope. The Editor-in-Chief has oversight of content and

the scope of the research accepted. The Editor-in-Chief is also responsible for ensuring quality control of the journal – creating and enforcing policies on rigour and process. Different types of editors may be used to handle the submitted research articles, any potential News and Views or summary articles, or perspective or policy pieces, among others. Sitting underneath this leadership are several associate and/or reviewing editors, covering more specialised topics. These editors are responsible for selecting peer reviewers and managing the peer review process with the editors, the publication company and the reviewers. Decisions about publication are made at the highest level, whereby the editor in chief bears responsibility, but recommendations are made by the editors at lower levels. *Figure created in BioRender.*

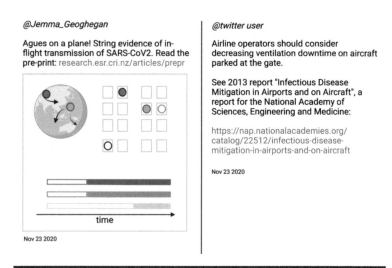

Figure 8.2 Communicating Results via Social Media. These can often lead to refinement of science ideas or creation of new ones. The figure shows a representation of a Twitter conversation after a paper was published showing transmission of

SARS-CoV-2 on a flight, studied using genomic sequencing. The ongoing discussion included public health measures (masks), business and economic interests and led to an example of a scientific report in a different discipline. Figure created in BioRender, and adapted from original Twitter screenshots. @ Jemma_Geoghegan tweet used with permission.

also solicit articles directly by emailing researchers inviting research articles, but this practice can also occur for genuine journals. It can be difficult to distinguish good open access journals from so-called predatory publishers and for new researchers. It can be easy to mistake a predatory journal for a high-quality journal. Peer review policies and information on editorial boards are usually available on journal websites and often a simple email to the editor can determine whether the journal is legitimate or not. There are also lists of predatory journals or publishers online that are openly available, although these lists are subject to their own biases.

Summary

The importance of publishing scientific results is not in dispute; however, new technologies and changes in expectations for publishers and scientists mean that we are about to start an exciting new phase of dissemination of scientific research. This will require a massive change in mindset in determining quality of work as well as the researcher.

Example 8.1 Incorrect Data – Western Blots

Western blots measure protein expression in cells. Because many blots show multiple experimental parameters and controls, they can be quite cumbersome to include in a research article. Therefore, the bands in Western blots are often chopped out and recompiled to create a more easily interpretable. However, this practice means that mistakes can happen, with the wrong bands included in the figure and in a small number of cases, deliberate re-positioning of bands can happen. There are numerous examples of erroneous Western blots online (see also Chapter 6). Artificial Intelligence (AI) is emerging as an issue for data image manipulation. A recent article addresses how Western blot reporting can be improved – http//:doi.10.1371/journal.pbio.3001783

Example 8.2 Data Repositories

There are many examples of data repositories, some of the bigger ones are NCBI GEO in the United States (www.ncbi.nlm.nih.gov/geo/) and ENA (European Nucleotide Archive; www.ebi.ac.uk/ena/browser/) worldwide. These are searchable raw data repositories that contain genomic and transcriptomic sequences from a variety of species including *Arabidopsis thaliana*, *Mus musculus*, *C. elegans* and *Homo sapiens* among others. Searching these databases can be a daunting task, however, the wealth of information that can be gathered from these searches and repositories is worth it. For example, some key word searches can reveal whether the experiment you are proposing may have already been done, thus avoiding repetition. Alternatively, existing data may help you focus your research or, confirm that your hypothesis is novel.

Potential Careers

Scientific publishing is the major mechanism to communicate scientific knowledge. However, there are many steps along the way to the publication of new research. Scientific journals offer many different opportunities for science graduates with an interest in science communication.

Journal Commentary Editor

Many journals include summary and perspective commentaries on the published data in the journal. Usually, a dedicated editor identifies new research and invites researchers in that field to comment on the newly published research. These editor positions are essential in translating sometimes large and complex ideas to a more general audience. It is important to have a knowledge of the research area and to be able to identify people who could contribute to these perspective pieces. Good writing and communication skills are essential, and these roles are perfect for those science graduates who like to follow lots of types of research.

Resources and Further Reading

Resources Used in This Chapter

- A summary of altmetrics and how they work www.altmetric.com/about-altmetrics/what-are-alt metrics/

■ Research article demonstrating SARS-CoV2 transmission in an aeroplane

Swadi T, Geoghegan JL, Devine T, McElnay C, Sherwood J, Shoemack P, Ren X, Storey M, Jefferies S, Smit E, Hadfield J, Kenny A, Jelley L, Sporle A, McNeill A, Reynolds GE, Mouldey K, Lowe L, Sonder G, Drummond AJ, Huang S, Welch D, Holmes EC, French N, Simpson CR, de Ligt J. Genomic evidence of in-flight transmission of SARS-CoV-2 despite predeparture testing. *Emerg Infect Dis.* 2021 Mar;27(3):687–693. doi: 10.3201/eid2703.204714.

■ An overview of the peer review process within a journal structure
www.biomedcentral.com/getpublished/peer-rev iew-process
■ A discussion of the current practices in peer review

Kelly J, Sadeghieh T, Adeli K. Peer review in scientific publications: Benefits, critiques, and a survival guide. *EJIFCC.* 2014 Oct 24;25(3):227–43.

■ *BMC Research Notes*

This journal allows publication of negative data, as well as data that supports previously published work. The data may lack some novelty while still providing useful information for other researchers.

https://bmcresnotes.biomedcentral.com/
about?gclid=Cj0KCQiAofieBhDXARIsAHTTldoL0e0o
zT8Z5hBhTwpFIcMtev4HhyJvv6j-3glKHdh9LTXiN9dvS
nUaApE1EALw_wcB

■ A book for students that supports and teaches good writing skills

Williams JM. *Style: Ten Lessons in Clarity and Grace.* Longman, New York, 2003.

■ Data repository sites:
  www.ncbi.nlm.nih.gov/geo/
  www.ebi.ac.uk/ena

Additional Websites That Might Be Useful

■ A professional society (International Science Council) has published the seven principles for scientific publishing
  https://council.science/current/news/seven-pri nciples-for-scientific-publishing/
■ A volunteer organisation that tracks scientific retractions and their impact
  www. retractionwatch.com
■ Assessing image data manipulation
  www.nature.com/articles/d41586-020-01363-z
■ Recording peer review contributions
  https://publons.com/wos-op/
■ American Society of Microbiology journal editorial policies as an example
  https://journals.asm.org/editorial-policies

- List of *Nature* portfolio journals as an example of diverse range
  www.springernature.com/gp/librarians/products/journals/nature-research-journals
- Frontiers peer review platform
  www.youtube.com/watch?v=qxnZ2zXCKv0
- A discussion of predatory journals
  www.nature.com/articles/d41586-019-03759-y
- The importance of publishing negative data

Nimpf S, Keays DA. Why (and how) we should publish negative data. *EMBO Rep.* 2020 Jan 7;21(1):e49775. doi: 10.15252/embr.201949775.

Research Publications for Further Reading

1. Discussion of how to improve Western blots for publication

Kroon C, Breuer L, Jones L, An J, Akan A, Mohamed Ali EA, Busch F, Fislage M, Ghosh B, Hellrigel-Holderbaum M, Kazezian V, Koppold A, Moreira Restrepo CA, Riedel N, Scherschinski L, Urrutia Gonzalez FR, Weissgerber TL. Blind spots on Western blots: Assessment of common problems in western blot figures and methods reporting with recommendations to improve them. *PLoS Biol.* 2022 Sep 12;20(9):e3001783. doi: 10.1371/journal.pbio.3001783.

2. Assessing fraud when reading or reviewing scientific articles

van der Heyden MAG. The 1-h fraud detection challenge. *Naunyn Schmiedebergs Arch Pharmacol.* 2021 Aug;394(8):1633–1640. doi: 10.1007/ s00210-021-02120-3.

3. Discussion of journal impact factors

Larivière V, Sugimoto CR. The journal impact factor: A brief history, critique, and discussion of adverse effects. In: Glänzel W, Moed HF, Schmoch U, Thelwall M (eds) *Springer Handbook of Science and Technology Indicators.* Springer Handbooks. Springer, Cham, 2019. https://doi.org/10.1007/978-3-030-02511-3_1

4. Discussion of Open Science policies and issues for some research

Besançon L, Peiffer-Smadja N, Segalas C, et al. Open science saves lives: lessons from the COVID-19 pandemic. *BMC Med Res Methodol.* 2021;21:117. https://doi.org/10.1186/s12874-021-01304-y

Exercises

Individual

Exercise 1

Find a research article that has also been promoted on social media. Which platforms have the authors used? What new ideas or information was provided by others commenting? What are the advantages and disadvantages of taking this approach.

Exercise 2

Visit the website: Retractionwatch.com. Choose a recent article and discuss why it was retracted. What could the authors have done differently? What is the effect of citations after retraction?

Class

Exercise 3

Take a research article and remove the abstract. Switch with another student and read their paper. Then write an abstract.

Exercise 4

Should we pay for peer review? Imagine you are part of the editorial team at Nature Publishing Group and there is a proposal from the editors to compensate reviewers for their time in reviewing manuscripts.

Prepare a list of pros and cons for a compensated review system and discuss the controversy. Why might scientists appreciate compensation for their time? Why is there pushback from publishers?

Some resources discussing payment for peer review:

■ Scientists, Publishers Debate Paychecks for Peer Reviewers
www.the-scientist.com/careers/scientists-pub lishers-debate-paychecks-for-peer-reviewers-68101?utm_campaign=TS_DAILY%20NEWSLETT ER_2020&utm_medium=email&_hsmi=100797 937&_hsenc=p2ANqtz-8C4zOBRw0I2Nogk2VJL goz-2NhWh63YCOh94vmkyFBOtfV0hKZwSoty-vvPEJCWkhIFQXZzBxPPdf9KWdLwXidg1o370ku Spoz3uZbLL54UZbh2P0&utm_content=100797 937&utm_source=hs_email

■The $450 Question: Should Journals Pay Peer Reviewers
www.science.org/content/article/450-question-should-journals-pay-peer-reviewers

Chapter 9

Bias

Introduction and Scope

In this chapter, we address a common problem in science – bias – and hopefully identify ways for students to detect and resolve the problem. Bias refers to the interpretation or analysis of information that is not completely fair. Everybody experiences bias, whether deliberately or not, but acknowledging its existence is most important. Bias reflects an individual's upbringing, culture, training and exposure to media or other science opinions. It is a complicated concept but the problems it causes are surmountable. There is extensive literature on bias in general, written by experts (which we are not), as well as on the effect of bias on science; here we provide a primer on bias and how it affects science decisions.

 DOI: 10.1201/9781003326366-9

Learning Objectives

- To identify different types of bias in experimental research and compare to biases that researchers themselves hold
- To discuss possible solutions to experimental bias when designing studies.
- To determine how subconscious bias may affect your career choices and opportunities

Concepts of Bias

Confirmation Bias in Science

Confirmation bias is the idea that you have predetermined your experimental result. For example, you know that protein A will lead to a higher readout than Protein B. This is very difficult to negotiate, as hypothesis-directed research can require a predetermined result that the researcher is trying to prove. Ideally, proving a null hypothesis is a better option, that is, showing that you can exclude a particular idea means that you are still effectively keeping your options open. The other alternative is to create a more vague hypothesis, for example, the intervention will lead to a change in protein amount, rather than an increase or decrease.

However, confirmation bias can have long-term effects on both an individual's science and the scientific community. Established dogma can be difficult to overcome, especially when there is good evidence to support it. The danger of confirmation bias is that readers and scientists are reluctant to believe research that contradicts dogma, even if it is of high quality. Further, it can be difficult to obtain funding to disprove established dogma, as those peer reviewers assessing the proposal may believe the dogma. However, scientific research is about challenging facts and revisiting ideas. Therefore, while confirmation bias exists, most people are still open to new ideas. Further, confirmation bias is easy to acknowledge and, in many cases, to include in research design or interpretation, and does not necessarily come with values attached.

Subconscious Bias

Subconscious bias is quite different. There are many articles from specialists in this area, including subconscious bias based on, for example, gender or race, in all aspects of society. Again, we try to keep the focus on the relevance to biomedical research. Most people have some form of subconscious bias, and so it is a common problem. Subconscious bias refers to inbuilt prejudices or opinions of individuals. These may be generated through a lifetime of experience or coaching but can have significant effects on the quality and type of scientific research that is performed and published. Subconscious bias or perhaps the seriousness of the impact it can have is usually not immediately recognised by an individual.

In science, examples of subconscious bias include assumptions that women perform worse research than men, that black people are less able to lead a project, or that women are better at perceived soft sciences, such as psychology, than men. These are extreme examples and, in reality, the effects and decisions are much more subtle. Stereotype bias is effectively a subset of subconscious bias and refers to decisions made based or reinforced by stereotypes we see in society. There are vast numbers of scientific studies proving the existence of subconscious bias (see Resources).

Experimental Bias

Experimental bias refers to the smaller picture – how decisions we make in experimental design and data analysis can influence the results that are obtained, or the interpretation of those results. When we design experiments, we obviously run the risk of confirmation bias – if we expect to see a change in the level of one cytokine, we may choose to measure only that cytokine and not any others. But then how do we know there are not differences in things we don't measure? But how can we possibly measure everything? The solution is to measure what you can, but to acknowledge in the text the limitations of not studying other things. As a reviewer, you can request that scientists add extra parameters or groups, or to include more discussion on limitations.

A very good example of experimental bias is the choice of animals in experiments. Researchers often use mice because they are available, reagents have been optimised for this model and techniques are well established in mouse models. But is it the best model? Why do you think so? Because it is all you have ever seen? Are some of these experiments actually feasible in clinical studies? Do they add to the generation of new data? These are the questions that researchers must be prepared to consider and ideally address.

BOX 9.1   SEX BIAS IN EXPERIMENTS

An interesting development in biomedical research is the effect of sex on experimental results. Traditionally, human males have been used over 100 years to research biology, primarily because women are seen to be more complicated, especially when including pregnancy. Further, populations of homogenous humans are more likely to be able to be found and grouped together, such as the all-white all-male Doctor's Study (www.ctsu.ox.ac.uk/research/british-doctors-study) that famously identified smoking as a cancer risk. However, this has created an industry of medical research focused entirely on the biology of white men, and even the 70 kg average white man. Given the diversity of actual patients, especially when at least half of them are not even men, this has been problematic in actually treating patients. Fundamental research has also been affected by sex bias – experimenters tend to use either male or female mice to keep populations homogenous. This also reduces variability, which is good science, and animal numbers, which is good ethically. However, it has become apparent that many findings in male mice are not reproducible in female mice or vice versa, and that sex differences affect not only the obvious areas of endocrinology or reproduction, but also neuroscience, immunology and even microbiology. To this end, the NIH now requires grant applications to study research questions in both sexes of mice.

Impact of Bias

The impact of bias in both experiments and greater science can be massive. There are several areas of bias that are worth highlighting.

Effect of Gender

There have been numerous studies showing that female scientists are less likely to have papers accepted in journals or as talks at conferences. Retrospective blinding studies have shown that the less representation of women is due to bias against women. This is obviously more complicated by numerous factors, but it means that women are underrepresented in science and that this is not changing quickly.

Effect of "Big Name" Labs

The nature of scientific research means that success breeds success and "big name" labs can become established and have a major role in determining the direction of scientific research. Often this is based on the high-quality research from these labs, which leads to more funding, more publications and even more quality research. However, small labs run by non-famous people can also do high-quality research, but this may not be noticed, or be published in a high-impact journal. Blinding of review is one way to reduce the effect of big name bias. For example, to be invited to present at a conference is a big honour for the researcher, and also a drawcard for the organisers to recruit more people to pay to attend the conference. Do you then want to stack your line up with people that more scientists will recognise as famous (for example, by having lots of Nobel prize winners)? Or do you find new innovative researchers with a lower profile but who may have brand new exciting ideas? Because of the history of bias, the big-name bias also reinforces both gender and racial/country bias too.

Cultural Bias

What we have termed cultural bias is a mixture of other biases. We really mean here that the way people interpret scientific results or findings is influenced by what we already believe and how we choose to interpret data. Biological science has been dominated by Western (usually white) world views, in terms of what is considered important and in how experiments are designed. Even this entire book has been hugely influenced by our training in a white Western male-dominant view. Hence, the idea that we compete for funding and measure our success by research articles reflects only one very narrow world view of progress or success. For example, societies and cultures that have a more community-wide view of research impact may wonder why so much emphasis is placed on the discovery of one molecule changing the behaviour of one cell.

Consider the anti-vaccine movement. As immunologists, we find it very frustrating that people choose to ignore the scientific evidence in favour of vaccination. However, if our argument is predicated on a belief that something is true if the scientific evidence proves an effect, but the counterargument comes from someone unused to working in a framework of claim and evidence, and the ability to critically analyse scientific evidence, then the argument quickly reaches an impasse.

BOX 9.2   PROXIMITY BIAS

This bias is a new concept, which was highlighted during the COVID-19 pandemic and the associated lockdowns. The ability to work from home during lockdowns varied hugely, with many research articles showing a negative impact on some sectors. However, the increased acceptance of hybrid work environments, inadvertently tested during lockdowns, has generally been seen as a positive outcome. Employers, including science and research employers, are more likely to approve flexible work options now than prior to 2020. The choice to go into work versus working from home, however, is not as straightforward as it may seem.

There are biases that are apparent when some employees are more visible than others. This was recently highlighted in this commentary – www.bbc.com/worklife/article/20210804-hybrid-work-how-proximity-bias-can-lead-to-favouritism. The commentary refers to a known phenomenon called proximity bias, where individuals who are seen or more commonly interact with their managers can be seen as more competent or more valuable than those that are seen less often. Proximity bias has existed pre-COVID-19, but employers must consider this effect when judging performance of those in a flexible work format. Several strategies are highlighted in the commentary to help reduce the effect of proximity bias.

Solutions to Bias

Fortunately, there are small things that we can do to address bias in all ways in science. That is not to say we can fix it, but there are three key areas to consider.

The first is being aware that bias exists. It is a good thought exercise to see if you can identify your own subconscious, stereotype biases and then consider why you may hold these biases. What biases might you have? Do you hold your male lecturers in higher regard than your female ones? Do you feel more comfortable asking a female lecturer a question than a male one? Experimental biases can be found when reading papers and discussing them in journal clubs – what parameters were measured? Is there any confirmation bias in the results? Being aware of your biases and those of other people is the first step to reinterpreting the science.

The second consideration is acknowledgement. Knowing that biases exist is fine, but when presenting data, it can be very important to at least acknowledge publicly that the work may be affected by your own bias. Including a discussion of potential interpretations if hypothetical biases are removed can only improve the quality of the science. For example, including a sentence in the text that the experiment was only performed on young mice, and commenting that results may be different in old mice, at least gets the readers to think about it too, or may even provide the opportunity for researchers in the same field to then design their experiment using old mice to complement your work and to advance scientific progress more effectively.

The final consideration is communication and discussion. From a personal development perspective, becoming aware of bias and acknowledging it is important but as a science community, these conversations need to be held at a higher level. Ask questions in a seminar that highlights bias. If you go to a conference, challenge the organisers about the lack of women speakers or speakers of colour, or why all the speakers are from Europe or United States. Support journals that require a research impact statement that provides information to a general audience about why the research was important.

Figure 9.1   Draw a Scientist. Scientists drawn by a male six-year-old (Walter Howard-Smith, left) and a female six-year-old (Maia Caswell, right) "Fifty years of drawing a scientist" asked school-aged children to draw their depiction of a scientist and asked to depict the scientist doing their job. In the 1960s and 1970s, almost all students drew a scientist as male, with glasses, facial hair and lab coat surrounded by laboratory equipment. By 2015, the percentage of students drawing male scientists dropped to about 65% with female student depicting female scientists 58% of the time. Still, male students depicted male scientists about 80% of the time. Surprisingly, the depiction of male scientists by both male and female students increased with age such that by high school, 75% of female students and 98% of male students depicted scientists as male. This is despite the fact that women earn (on average) approximately 45% of PhD degrees in STEM (science, technology, engineering and mathematics) fields (including Engineering and Physical Sciences). Young female students may not be making the connection that they too can become a scientist and an implicit bias remains. Here is link to the article: https://srcd.onlinelibrary. wiley.com/doi/full/10.1111/cdev.13039

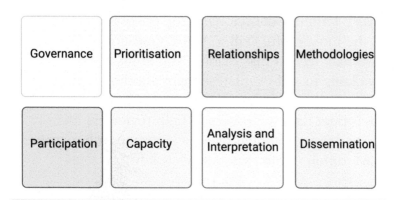

Figure 9.2 Colonial Cultures and Indigenous People in Research. The progress of biomedical research, and especially clinical research, has suffered in those places where indigenous populations have been excluded through bias, in research and in research priorities. Inclusion of a wide range of views and approaches leads to higher quality research. In 2019, a publication from New Zealand proposed the CONSIDER statement – consolidated criteria for strengthening reporting of health research involving indigenous peoples. There are 17 principles organised into eight topics – all or some of which can be used by researchers when reporting their work. The CONSIDER statement aims to strengthen both research and health outcomes for indigenous people. *Figure created in BioRender.* Huria T, Palmer SC, Pitama S, et al. Consolidated criteria for strengthening reporting of health research involving indigenous peoples: the CONSIDER statement. *BMC Med Res Methodol.* 2019;19:173. https://doi.org/10.1186/s12874-019-0815-8

Summary

Like ethics, bias is a complicated concept, and one that is hard to address for an individual but also for an institution or discipline. An awareness of potential bias ("I am not conscious of my unconscious bias") is the most useful skill to learn. Question whether you may have a bias or not. Numerous resources are available to self-evaluate – many funding bodies or journals include these self-review exercises as part of their peer review process to help reviewers identify and address potential bias. Being open to how others have different life experiences is key.

Example 9.1   Increasing Diversity in the Publication Process

Many journals have commenced diversity projects to increase the type of people reviewing, writing, publishing and editing publications, in order to reduce some of the biases described in this chapter. The simplest way to do this is to record the data of individuals involved in the processes to determine where there are gaps or disproportionate representation. Some of these include sex (including gender diverse), race, country of origin, country where the research took place and budget of research laboratory. Below is a list of the many ways publishers and journals attempt to increase diversity in science.

Collecting diversity data from:

- Authors
- Reviewers
- Editorial board members

Ensure people understand unconscious bias

- Send training tools to reviewers of manuscripts and grants

- Require completion of such training to perform the task
- Provide ways for researchers to show they have completed training

Publish research for diverse, but small, audiences

- Many high impact, and therefore perceived good journals, many not see the wide impact of research on a topic relevant for a small country, a rare disease or an indigenous population

Provide financial support for researchers in need

- Publishing is expensive and many researchers do not have budgets for publishing costs or open access costs
- Journal publishing companies often offer financial support or fee waivers to under-resourced researchers

Example 9.2   Measuring Research Performance and Including Bias

In a recent article studying awards of independent grant funding, the authors found that black researchers were less likely than white, Asian or Hispanic researchers to be awarded funding. However, this could be partly corrected by applying the productivity metrics of a successful applicant to a black applicant pool. This suggests that the (traditional research) productivity of black people was a factor in the awards – including the fact that research from black authors was less likely to be cited. This work points to a sequential and institutionalised racism that undervalues contributions from some populations, and that this has long-term effects.

Here is a link to the article: https://journals.plos.org/plosone/article?id=10.1371/journal.pone.0205929.

A similar study was performed in New Zealand, looking at the cycle of government funding provided to research institutions, and based on the metrics of its researchers. In this study, comparing the effect of gender on research value, the authors found that men were valued more highly than women, despite similar metrics. This was represented in promotion levels and pay scales and that breaks for children or differences in age did not explain the difference in value. "Indeed women whose research career trajectories resemble men's still get paid less than men. From 2003–12, women at many ranks improved their research scores by more than men, but moved up the academic ranks more slowly."

Here is a link to the article: https://journals.plos.org/plosone/article?id=10.1371/journal.pone.0226392

Potential Careers

Bias exists in all forms of science, and there are many types of bias, both experimental and social. One of the most difficult jobs in science is sharing bias data and analysis with scientists who are not expert in this field or in these types of analyses. Science communication is a massive field and there are huge opportunities in communicating science results as well as science method to the public, and to non-specialists. This includes sharing information on bias – possibly one of the most important aspects of research.

Science Communication

Many institutions offer postgraduate science communication degrees – these are qualifications that train a student with an undergraduate science degree in forms of science communication; however, many successful science communicators have drifted there because of a desire to share their own research. Some aspects of science communication include film making (for example, documentary films), writing (for example, medical writers), blogging and media coverage (for example, social media blogs) and science infographics (for example, for patients). Science communicators played an invaluable role during the COVID-19 pandemic, translating complicated scientific concepts into readily understandable actions, for the public and for policymakers.

Resources and Further Reading

Resources Used in This Chapter

- Proximity bias
  www.bbc.com/worklife/article/20210804-hybrid-work-how-proximity-bias-can-lead-to-favouritism
- Effect of colonial cultures

Hood AM, Booker SQ, Morais CA, Goodin BR, Letzen JE, Campbell LC, Merriwether EN, Aroke EN, Campbell CM, Mathur VA, Janevic MR. Confronting racism in all forms of pain research: A shared commitment for engagement, diversity, and dissemination. *J Pain*. 2022 Jun;23(6):913–928. doi: 10.1016/j.jpain.2022.01.008.

■ Nature Publishing Group plan to increase diversity

Else H, Perkel JM. The giant plan to track diversity in research journals. *Nature*. 2022 Feb;602(7898):566–570. doi: 10.1038/d41586-022-00426-7.

■ Draw a Scientist

Miller DI, Nolla KM, Eagly AH, Uttal DH. The development of children's gender-science stereotypes: A meta-analysis of 5 decades of U.S. draw-a-scientist studies. *Child Dev*. 2018 Nov;89(6):1943–1955. doi: 10.1111/cdev.13039.

■ Research performance metrics

Ginther DK, Basner J, Jensen U, Schnell J, Kington R, Schaffer WT. Publications as predictors of racial and ethnic differences in NIH research awards. *PLoS One*. 2018 Nov 14;13(11):e0205929. doi: 10.1371/journal.pone.0205929.
Brower A, James A. Research performance and age explain less than half of the gender pay gap in New Zealand universities. *PLoS One*. 2020 Jan 22;15(1):e0226392. doi: 10.1371/journal.pone.0226392.

■ The CONSIDER statement

Huria T, Palmer SC, Pitama S, et al. Consolidated criteria for strengthening reporting of health research involving indigenous peoples: the CONSIDER statement. *BMC Med Res Methodol.* 2019;19:173. https://doi.org/10.1186/s12874-019-0815-8

Additional Websites that Might Be Useful

■ The Doctor's Study
www.ctsu.ox.ac.uk/research/british-doctors-study
■ An example of a professional society's plan to address equity, diversity and inclusion
www.immunology.org.au/asi-programs-and-opportunities/womens-initiative/asi-gender-equity-and-inclusion-policy/
■ Sexism in academia
www.nplusonemag.com/issue-34/essays/sexism-in-the-academy/
■ Unconscious bias in medicine
www.youtube.com/watch?v=pMig1kHtufo
■ Understanding unconscious bias
www.npr.org/transcripts/891140598
■ NIH policy to reduce bias in grants to big name institutions
www.nature.com/articles/d41586-022-04385-x?WT.ec_id=NATURE-20221222&utm_source=nature_etoc&utm_medium=email&utm_campaign=20221222&sap-outbound-id=C865AD31EBED8C03F196420F44C8DE4875A0C488
■ Website for companies to provide help in reducing bias in all activities
https://biasinterrupters.org

■ Perspectives on gender bias in science from a transgender scientist https://med.stanford.edu/news/all-news/2006/07/ transgender-experience-led-stanford-scientist-to- critique-gender-difference.html

Research Publications for Further Reading

1. Gender inequity and consequences

Borger JG, Purton LE. Gender inequities in medical research funding is driving an exodus of women from Australian STEMM academia. *Immunol Cell Biol.* 2022 Oct;100(9):674-678. doi: 10.1111/imcb.12568.

2. Experimental bias (microscopy)

Jost AP, Waters JC. Designing a rigorous microscopy experiment: Validating methods and avoiding bias. *J Cell Biol.* 2019 May 6;218(5):1452–1466. doi: 10.1083/ jcb.201812109.

3. The case for using both sexes in animal experiments

Miller LR, Marks C, Becker JB, Hurn PD, Chen WJ, Woodruff T, McCarthy MM, Sohrabji F, Schiebinger L, Wetherington CL, Makris S, Arnold AP, Einstein G, Miller VM, Sandberg K, Maier S, Cornelison TL, Clayton JA. Considering sex as a biological variable in preclinical research. *FASEB J.* 2017 Jan;31(1):29–34. doi: 10.1096/fj.201600781R.

Diester CM, Banks ML, Neigh GN, Negus SS. Experimental design and analysis for consideration of sex as a biological variable. *Neuropsychopharmacology*. 2019 Dec;44(13):2159–2162. doi: 10.1038/s41386-019-0458-9.

Exercises

Individual

Exercise 1

You are organising a scientific conference for your national professional society. What should you consider when inviting speakers? How will you ensure diversity?

Exercise 2

Consider a research paper and identify any sources of experimental confirmation bias. Rewrite the text, with the same data, but include acknowledgement of any experimental biases and how they may affect the interpretation of results

Class

Exercise 3

Choose a disease that has been extensively researched. Assess the number of male versus female participants (animal or human). Is it even? Are both sexes included in each study? What are the long-term effects of decisions made based on this research?

Exercise 4

Find a high-profile research paper (for example, high impact or many citations) that explains a biomedical phenomenon. Now hunt for other papers published before or after this paper that support or argue against the results found. Is there an effect of big name lab? How could you address it?

Chapter 10

Soft Skills in Science

Introduction and Scope

Science and research require a range of skills.
Researchers and scientists need to manage multiple
funding requests, write papers, teach and mentor
students and review other people's grants and papers.
Public outreach and science communication are also
part of a science career.

The upside of this is that scientists gain an
enormous and varied skill set – they must be creative,
motivated, organised, competent at managing budgets;
they must understand both conceptual and practical
aspects of science, write and speak well. This means
that science graduates are superbly placed for most
job or career opportunities, highlighted throughout
this book, regardless of whether the role requires
actual scientific techniques or prior knowledge of
a specific field. Above all, the ability to critically

DOI: 10.1201/9781003326366-10

evaluate data and decide if it provides evidence for a conclusion, is a vital life skill.

The downside of the multitasking nature of science as a career is that it can be difficult to juggle so many simultaneous tasks and be skilled in everything all at once. This is also true for those studying for a science degree. Students must manage multiple courses and assessments, and manage lab teaching, lectures, tutorials; often while also earning money in an external job. Many students also have significant family commitments. Stress and anxiety among all students are common. We are not expert social workers or psychologists; however, this chapter will provide some guidance on the soft skills of science – managing time and energy, prioritising tasks and remaining positive. Science is hugely rewarding but can have some big road bumps along the way.

Learning Objectives

- To identify the skills required for a science career
- To discuss how certain skills can prepare you for different careers in science
- To provide you with strategies to manage a busy study programme

Effective Multitasking

When studying for a science degree, undergraduate students often have a heavy contact hour workload – they not only need to fit in lectures and tutorials but also substantial hours of practical lab coursework. The lab work, like real science, can be fraught with failure and require repeats of experiments, or working in a group of varied ability. At the same time, assignments are due and exams are imminent. How does a student manage all these tasks? The good news is that the skills learnt to manage many things at once as a student directly helps in any science career – very few scientists have only one thing going on or are meeting one deadline at a time.

For effective multitasking, there are two key things to be aware of: (i) recognising what needs to be done and (ii) keeping track of what has and hasn't yet been done. Linked to these factors is the timeline of what needs to be done by when (see next section). Implicit in these ideas is the need to be organised, and to know what is coming and when. This means that all the paperwork and email contact from course administrators actually does need to be read and stored somewhere for easy access.

One successful mechanism for good multitasking is making lists. Everybody makes their own lists their own ways, and all of us have enjoyed the satisfaction of crossing off tasks on a list (including Task 1: make a list).

An example of a successful strategy is to create multiple lists for various tasks:

■ Tasks for this week – including a class timetable, assessments etc. These are things that can be fitted into the timetable for this week.

- Tasks for next week – things that are directly coming up, but also things that you may need to prepare for, such as the next assessment.
- A list of tasks that you must do at some point and not forget about.

Each week some of the tasks can move to this week, next week or remain on the eventually list. The most successful strategy is to put absolutely everything on one of the lists. Once it is recorded, it can come out of your brain, and you don't have to worry about remembering that at some point you have to do it–it's on the list and now all you must do is organise it.

It is important to recognise that good multitasking is something you learn and if you are studying towards a science degree, it is part of the skillset that you acquire but it may not be part of the skillset you currently have. Recognising that you aren't necessarily great at it shouldn't be an issue – it is fine to ask for help in planning and keeping track of tasks. Everybody around you has strategies for staying on top of workloads and is happy to share their strategies. Peer support groups allow students to learn strategies from each other, particularly when students have the same or similar workloads. A potential employer will always ask about strengths and skills and being able to demonstrate how many classes and assignments you managed, and to explain your strategy for managing them, will be considered a strength.

Working in Teams

Science is a collaborative process, and as a scientist you will constantly be working with other people. As a science student, you will almost certainly work in groups for aspects of your study, such as a group lab project or a group assessment. A key skill in science is to discover how to work with other people. We have all been in groups where a dominant voice dictates what happens – sometimes that is us, other times, we are the quiet one. Communicating well is a difficult group skill to learn but is essential for later science career decisions and progress.

For effective group work, the first meeting is the most important. In this meeting, establishing a division of workload is key, playing to each person's strengths, but also allowing the opportunity to learn as well. A timetable is also very helpful – by what time should each action be completed? A regular meeting time is useful to make sure everybody is on track, rather than hoping everybody will get everything done by the deadline. It is sometimes useful to have a group "pact" at that first meeting that outlines everyone's workload and due dates, especially when your group work involves a grade for an assignment or a firm deadline for a funding agency. The pact can also outline consequences when members of the group do not hold up their end of the bargain. This is useful when communicating to instructors the amount of work each student is responsible for. When writing this book, the two authors were based in different countries, in two different time zones, and we met every two weeks at the same time to discuss progress and plan the tasks for the following week. This kept us on track, on time, and productive.

Time Management

The ability to manage multiple tasks *on time* is a vital life skill and something that takes time to learn. A science degree structure, along with the multitasking strategies above, supports the development of time management as a real and provable skill. Scrambling from one assignment or exam to the next is a bad idea for your grades, but the structure of a degree means that you can take advantage of multiple and overlapping deadlines to develop the skills needed to manage multiple projects – a skill that can be demonstrated in a job interview.

Time management has two facets – one is the deadline and the other is the time required to actually complete the task. For each new task, it is essential to know the deadline (check it multiple times!) and to know how much time it is going to take to achieve the goal. Working out the hours required to complete it, especially if these hours are dependent on multiple factors, allows you to create a realistic and achievable timetable. Students often plan when they will start and complete a task without actually considering the workload that goes into it. The workload also depends on your own strengths and weaknesses. If you read data quickly or can write very well, then those aspects of the work will take you less time than someone who still needs to develop writing skills. This can be especially problematic for students learning in their second or third language.

A realistic timetable also considers external events. If you are taking five days off to visit your family, then that needs to be included in the timetable as days that are non-productive, and therefore more work needs to be fitted in before or after the holiday. Holidays can always be included, as long as the planning has been done in advance. Similarly, if students are managing study as well as external work, then trying to complete difficult tasks after working a 12-hour shift is not going to go well. Instead, save the important but less difficult tasks (for example, organising data into files) for those times, and keep the hard thinking tasks for days when you are rested and don't have other activities. Allow time for proofreading and time for formatting issues to be resolved (there will always be formatting issues). Group work, as discussed earlier, also needs to be included in the timetable. Setting realistic goals (see later) is an important part of time management.

Energy Management

Energy management and time management are closely intertwined. In this context, energy management is about defining what gives you energy and what takes it away, and how to use that knowledge to set goals, plan time management, and to manage multiple tasks.

Recognising that at some times of day you are better suited to some kinds of tasks means that you can plan your tasks to take advantage of periods of high energy and avoid periods of low energy. If some things, for example, lectures, take your energy away, then it is wise to avoid planning important, high-level thinking tasks, immediately after lectures; instead, find a time when you do have more energy. Alternatively, sandwiching hard jobs between fun things can be a good way to maintain high energy, and not lose it to the hard work.

An important aspect of good productivity and good work is to maintain a "work–life balance". When considering what aspects promote high energy or low energy states, be sure to include external activities too. If socialising with friends makes you feel good, then do it, but plan when the work will be done as well. Don't go out the night before an assignment is due and then try to write the assignment at 6 a.m. the next day. Instead, plan the night out and allocate time to have the assignment completed before you go, then you will have more fun anyway.

One common mistake for students and for scientists is to persist through a low-energy state and try to complete work. Students often hit a roadblock and lose creativity or perform experiments badly when they are tired or not able to concentrate. Recognising that your energy is low and that you are unlikely to be productive is really important. It is much better to stop a task (for example, writing) if it takes three hours and all your energy to complete something small that should take 30 minutes. Clearly it is not a good use of time or energy, and you are better to step away and return to it later.

However, the greatest source of motivation is achievement. Anything that leads to the progression of a task will help to motivate you to carry on. A strategy that can be useful is to maintain a list of low energy tasks that still move a project along, for example, organising your lecture notes, or proofreading an assignment. These take much less time than planning a complicated experiment. These small but important tasks can be slotted in during periods of low energy to ensure that projects still move along, and that you are achieving something useful. Even though these seem like small tasks, it's still important to write them down to maintain that sense of progress.

It is a good idea to do a self-audit to define what things help you gain or lose energy. Then you are aware of these as they happen, but you can also plan around these. One of our students was a DJ in his spare time and plans to keep music as part of his career path. However, the late nights meant his study suffered. By pulling apart the music aspects of his life (performing is an energy boost, but staying up late is an energy drain), we could prioritise his difficult assignments to begin two days after a gig, but at a higher work rate than if the work had been spread throughout the week (for example, five hours a day of study, not three). This allowed him time to concentrate on both roles and maximise the outputs for each. It also stopped him worrying all weekend about when he would fit the study in – we just stopped him studying during the weekend. The study timetable each week had enough time for him to complete the work, it just needed to be arranged to suit his life.

BOX 10.1 MANAGING FAILURE AND REMAINING POSITIVE

Inherent in scientific experimentation and discovery is the chance that the experiment fails, or the protocol you are developing needs a lot of trial and error or troubleshooting to get up and running. These bumps in the road can wreak havoc on your carefully constructed lists, goals, and time management strategies. Add to that any infrastructure-related issues (loss of electricity, water, construction and repairs, a global pandemic) and often your work does not go to plan. Moving experiments or restructuring experiments due to circumstances out of your control may be difficult to handle. But you are ultimately in control of your own experiments and can create a solution. Having a positive, but realistic mindset about your experiments and your work is essential.

Choosing Priorities

At any one time, a scientist or a student has multiple tasks with different timeframes for completion. It can be difficult to keep track of all these tasks and to allow enough time to complete them to a high standard. Two concepts can help with this. One is the idea of Urgent/Important and the other is the idea of Quality.

Urgent / Important

This is a commonly used strategy for all sectors of society but is equally valuable in science. What is important is different to what is urgent. Assigning an urgency and importance value to all tasks can help prioritise. An upcoming exam is urgent and important, so it needs to be a high priority. An assignment due in four weeks is important but not urgent, so can be worked on but not as the sole project. Things that are not urgent or important (clean the fridge) can move down the priority list.

Quality

We are taught through the education process to aim high and make everything as good as it can be. However, when managing multiple projects, it is worth thinking about the time and energy taken to take an assignment from 95% to 96%. Sometimes, this incremental increase in grade is essential, for example, for entry into a competitive course, but often it makes no difference at all to immediate or long-term goals. So, consider the time being spent on making this 1% increase in quality – if it will take three hours and you have several other assignments due, then it is not worth the time to increase the grade. Similarly, if there are two assignments due on the same day, one worth 5% and one worth 25%, it is better to invest the energy in the 25% assignment.

The skill of assigning importance and quality to tasks can be difficult to adjust to after a lifetime of trying to do the best in every scenario. However, sometimes it is just not possible. Realise that *very good* and *very very good* are not that different. Detaching your own self-worth from the high achievement metrics is a very handy life skill.

Summary

Managing a science degree can be difficult – balancing all the aspects of study (or job) against time and energy limitations is a key skill for scientists. It is important to celebrate the vast array of skills you can acquire as a scientist and appreciate how valuable those skills are. It is also important to understand your own limitations and strengths – science is one of the most flexible careers possible – you will be able to work in a way that best suits you, so be sure you know what that is.

BOX 10.2   MENTAL HEALTH
AND SEEKING HELP

Sometimes we can't manage everything and sometimes we need to ask for help. Every university has support systems for students who are struggling with workload and the intersection between study and life. It is important to seek help if you need it – often your classmates are ideal people to ask for support or advice, but your lecturers or tutors will always help too.

Figure 10.1 Skills That Scientists Have. These are important skills for many different disciplines and careers. *Figure created using WordClouds.com.*

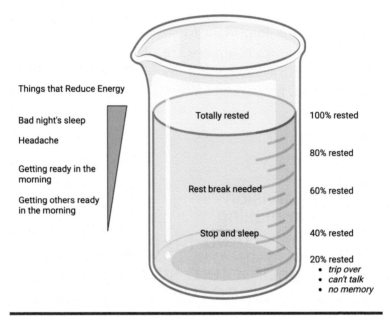

Figure 10.2 Energy Management. Every day we start fully rested but things can take away energy almost immediately. It is important to recognise when energy is getting dangerously low and plan to rest to recover. Adapted from Insight Rehabilitation/ ACC. *Figure created in BioRender.*

Example 10.1 Employability Skills

There are several websites that list skillsets specific for particular jobs, or for any job at all! Here is a list of seven skills from the careers.govt.nz website – a guide from the New Zealand government for students.

What are employers looking for? Most of these skills are ones that you learn as a scientist:

1. Positive attitude
2. Communication
3. Teamwork

4. Self-management
5. Willingness to learn
6. Thinking skills (problem solving and decision making)

Example 10.2   Exercise and Performance

Exercise Reduces Stress

Exercise leads to increases in the feel-good neurotransmitters, such as serotonin, dopamine and endorphins. Those who have experienced runner's high know how powerful the effect can be.

von Haaren B, et al. Reduced emotional stress reactivity to a real-life academic examination stressor in students participating in a 20-week aerobic exercise training: A randomised controlled trial using ambulatory assessment. *Psychol Sport Exercise.* 2015;20:67–75.

Exercise Improves Focus

Research has shown that bursts of exercise activate the parts of the brain that improve or shift focus. Important exam? Go for a run first. The effects can last for up to two hours.

Basso JC, et al. Acute exercise improves prefrontal cortex but not hippocampal function in healthy adults. *J Int Neuropsycol Soc.* 2015;21(10):791–801.

Exercise Improved Cognition

Many studies have shown that regular exercise improves our ability to learn new things and understand complex ideas.

Hogan CL, et al. Exercise holds immediate benefits for affect and cognition in younger and older adults. *Psychol Aging.* 2013;28(2):587–944.

Exercise Improves Mental Health

A recent meta-analysis of research papers studied the effect of exercise on depression, anxiety and distress. They showed exercise reduced these symptoms in both the general population as well as those with diagnosed mental health disorders.

Singh B, Olds T, Curtis R, et al. Effectiveness of physical activity interventions for improving depression, anxiety and distress: an overview of systematic reviews. *Br J Sports Med.* doi:10.1136/bjsports-2022-106195.

Potential Careers

The point of this chapter is to highlight how many skills scientists acquire during their training. It seems odd to pinpoint just one career that would select one of these diverse skills. Instead, we have listed some of the jobs that our undergraduate and graduate students have had that are outside the role of science. Sometimes parents ask students what job they plan to get with the major or degree that they have chosen – here are some of them:

- High-school science teacher
- IT manager
- Filmmaker
- Communications advisor
- Policy analyst
- Clinical trials manager
- Māori health advancement advisor
- Investment analyst
- Founder and CEO of data analysis company
- Government policy advisor

- Medical professional
- Research advisor
- Patent attorney
- Commercialisation manager
- Account manager
- Sales consultant
- Health, safety, and compliance manager
- Librarian
- Research and teaching IT
- Economist
- Marketing advisor
- Animal practice manager
- Scientific content manager
- Medical writer

Resources and Further Reading

Resources Used in This Chapter

- University of Otago, Dunedin, Academic Leadership Development Programme
- New Zealand Government www.careers.govt.nz
- Exercise, health and study www.theguardian.com/lifeandstyle/2023/mar/02/ exercise-is-even-more-effective-than-counselling- or-medication-for-depression-but-how-much-do- you-need?CMP=Share_iOSApp_Other

von Haaren B, et al. Reduced emotional stress reactivity to a real-life academic examination stressor in students participating in a 20-week aerobic exercise training: A randomised controlled trial using ambulatory assessment. *Psychol Sport Exercise.* 2015;20:67–75.

Basso JC, et al. Acute exercise improves prefrontal cortex but not hippocampal function in healthy adults. *J Int Neuropsycol Soc.* 2015;21(10):791–801.

Hogan CL, et al. Exercise holds immediate benefits for affect and cognition in younger and older adults. *Psychol Aging.* 2013;28(2):587–944.

Singh B, Olds T, Curtis R, et al. Effectiveness of physical activity interventions for improving depression, anxiety and distress: an overview of systematic reviews. *Br J Sports Med.* doi:10.1136/bjsports-2022-106195.

Additional Websites That Might Be Useful

- An opinion piece on the importance of soft skills in science
  https://blogs.scientificamerican.com/observati
  ons/soft-skills-in-the-life-sciences/
- A guide to soft skills
  www.lanl.gov/careers/diversity-inclusion/s3tem/
  index.php\
- A perspective on soft skills
  https://entomologytoday.org/2021/03/03/soft-ski
  lls-secret-ingredient-successful-science-career/

Research Publications for Further Reading

1. How a balanced life improves study

Feraco T, Resnati D, Fregonese D, Spoto A, Meneghetti C. Soft skills and extracurricular activities sustain motivation and self-regulated learning at school. *J Exp Educ.* 2022;90(3):550–569, doi: 10.1080/00220973.2021.1873090.

Feraco, T, Resnati, D, Fregonese, D, et al. An integrated model of school students' academic achievement and life satisfaction. Linking soft skills, extracurricular activities, self-regulated learning, motivation, and emotions. *Eur J Psychol Educ.* 2022. doi: 10.1007/s10212-022-00601-4.

2. The most useful soft skills

Jardim J, Pereira A, Vagos P, Direito I, Galinha S. The Soft Skills Inventory: Developmental procedures and psychometric analysis. *Psychol Rep.* 2022;125(1):620–648. doi: 10.1177/0033294120979933.

3. How to teach and learn soft skills in biological sciences

Beno SM, Tucker DC. Growing innovation and collaboration through assessment and feedback: A toolkit for assessing and developing students' soft skills in biological experimentation. In: Pelaez NJ, Gardner SM, Anderson TR (eds) *Trends in Teaching*

Experimentation in the Life Sciences. Contributions from Biology Education Research. Springer, Cham, 2022. doi: 10.1007/978-3-030-98592-9_20.

4. Communication is a key skill in science

Lee SH, Pandya RK, Hussain JS, Lau RJ, Brock Chambers EA, Geng A, Xiong Jin B, Zhou O, Wu T, Barr L, Junop M. Perceptions of using infographics for scientific communication on social media for COVID-19 topics: a survey study. *J Visual Commun Med*. 2022;45(2):105–113. doi: 10.1080/17453054.2021.2020625.

Exercises

Individual

Exercise 1

Perform your own assessment on your life and how you manage your time:

1. What gives you energy?
2. What takes energy away?
3. When are you most productive in terms of studying?
4. What would be the best time for you to sit an exam?

5. How would you design a study timetable to include exercise, socialising, eating and rest that would suit you?

Exercise 2

Create an urgent/important matrix for your current study workload. What things are urgent? What things are important?

Class

Exercise 3

Divide the class into multiple industries: government and policy, economics, healthcare, education, agriculture, manufacturing, food production, hospitality, transport, etc. In groups or as individuals, discuss the skills required to work in these industries and how a science degree could be useful.

Exercise 4

Throughout this book, we have asked you to work as a group within your class. Reflect on your role in these groups. Were you a leader? Were you outspoken? Did you do more work than others in the group? What strengths did you bring to the group? What skills would you like to improve for teamwork?

Index